ACTIVE NOISE CANCELLATION SYSTEM DESIGN ENGINEERING

OPTIMAL FEEDBACK CONTROL FORMULATION OF THE ACTIVE NOISE CANCELLATION PROBLEM: POINTWISE AND DISTRIBUTED

BY

K. C. ZANGI

RESEARCH LAB OF ELECTRONICS

MASSACHUSETTS INSTITUTE OF TECHNOLOGY

Wexford Press

2008

Contents

Chapter 1

Introduction

Unwanted noise is a by-product of many industrial processes and systems. In active noise cancellation (ANC), one introduces a secondary noise source to generate an acoustic field that interferes destructively with the unwanted noise, and thereby attenuates it.

Noise reduction is important to protect listeners in high noise environments from hearing damage, to enhance speech communication, and to reduce noise-induced fatigue. These adverse effects of noise can cause accidents and reduce the productivity of workers.

Passive silencers are low-pass acoustical filters commonly used on engines, blowers, compressor, fans and other industrial equipment. These silencers are very effective in attenuating high frequency noise; however, they are usually bulky and ineffective at low frequencies. This is because at low frequencies the acoustic wavelengths become large compared to the thickness of a typical passive silencer. A sound wave of 60Hz, for example, has a wavelength of 5.66 meters in air under normal conditions. It is also difficult to stop low frequency sound being transmitted from one space to another unless the intervening barrier is very heavy. Passive mufflers that are used to silence engine noise produce back-pressure by obstructing the turbulent flow out of the engine, and this back-pressure significantly reduces the efficiency of the engine. To overcome these problems, researchers have given active noise cancellation considerable attention recently.

The fundamental problem in ANC is to generate an acoustic field that interferes destructively with the unwanted noise field at the points of interest. A typical ANC system

13

utilizes several microphones to monitor the attenuated field and several canceling sources to generate the canceling field. The outputs of these microphones are used as inputs to some sort of electronic controller. This controller provides the inputs to the canceling sources in such a way that the acoustic field generated by these sources interferes destructively with the unwanted noise field at the points of interest.

Almost all existing ANC systems are designed explicitly to minimize the sum of the noise power at a finite number of spatial points. We shall refer to this type of performance criterion as a "pointwise performance criterion". Similarly, we refer to the resulting ANC system as a "pointwise active noise cancellation system". In pointwise ANC, it is commonly assumed that attenuating the noise at a finite number of spatial points will results in attenuation of the noise at all other points of interest. However, it has been shown in numerical studies that attenuation of noise at a finite number of spatial points can result in amplification of the noise at other points [15].

Existing controllers for pointwise ANC can be characterized as feedforward or feedback controllers. Adaptive feedforward controllers have been developed by a number of authors, and these controllers are designed to take advantage of the statistics of the signals involved. However, most existing feedback ANC systems are designed deterministically and do not take advantage of the statistical characteristics of the noise.

A distributed active noise cancellation system is defined as an ANC system that minimizes a distributed performance criterion. For example, a distributed ANC system might minimize the total acoustic energy in an enclosure. To the best of our knowledge, a mathematically rigorous procedure for designing distributed ANC systems has not been developed yet.

By formulating the ANC problem as an optimal feedback control problem, we develop a single approach for designing both pointwise and distributed ANC systems. The key strategy is to model the residual signal/field as the sum of the outputs of two linear systems. The unwanted noise signal/field is modeled as the output of a linear system driven by a white process. Similarly, the canceling signal/field is modeled as the output of another linear system driven by the control signal. Finally, the residual signal/field is modeled as the sum of the outputs of these two linear systems. We show that the control signal that minimizes a certain class of performance criteria is a linear feedback of the estimated states of these two linear systems. These state estimates are computed using a Kalman filter, and the feedback gain matrix/operator is obtained by iterating a Riccati equation. Note that in the pointwise case and in the distributed case, the control signal is not distributed and is used as input to an ordinary loudspeaker. Moreover in both these case, the control signal is generated based on the outputs of ordinary microphones that monitor the residual field.

While we focus specifically on the acoustic noise cancellation problem, the results developed in this thesis can be applied to other active cancellation problems. Vibration control is an example of a non-acoustic problem to which our results can be applied.

1.1   Outline of the Thesis

Chapter 2 motivates our study of active noise cancellation and presents a summary of the previous work done in this area. The physics of the problem are discussed in this chapter with emphasis on three dimensional acoustic fields. The different sound fields generated by ANC systems designed to achieve different acoustic objectives are studied. Several commonly used control strategies for ANC are presented, and the performance of these control strategies are analyzed and compared. The relationship between active noise cancellation and signal estimation is also discussed.

The optimal feedback controller for attenuating the acoustic pressure at a discrete number of microphone locations is developed in Chapter 3. Our approach is to model the residual signals at the microphones as the sum of the outputs of two finite-dimensional linear systems. The unwanted noise signals at the microphones are modeled as the outputs of a multiple-input/multiple-output linear system driven by a white process. Similarly, the canceling signals at the microphones are modeled as the outputs of a single-input/multiple-output linear system driven by the control signal. Finally, the residual signals at the microphones are modeled as the sum of the outputs of these two linear systems. We show that the control signal that minimizes a certain class of performance criteria is a linear feedback of the estimated states of these two linear systems. We also show that our formulation can be used to minimize the frequency weighted power of the residual signal. Such frequency dependent weighting might be important because of the difference in the sensitivity of the ear to sounds at different frequencies. These state estimates can be computed using a Kalman filter, and the feedback gain matrix can be computed by iterating a matrix Riccati equation. The single-microphone case is presented first, and the results are then extended to the multiple-microphone case.

In Chapter 4, several recursive/adaptive algorithms for modeling the unwanted noise at a single microphone are developed. We assume that the unwanted noise at the microphone is the output of an all-pole transfer function driven by a white process and develop a recursive/adaptive algorithm for estimating the parameters of this transfer function based on the measurements made by the microphone. An adaptive feedback ANC system can be obtained by combining the optimal control law of Chapter 3 with these recursive/adaptive algorithms for modeling the unwanted noise. Although we focus specifically on developing estimation algorithms for modeling the unwanted noise in the ANC problem, the algorithms presented in this chapter can be applied to the more general problem of identifying non-stationary autoregressive (AR) processes embedded in white noise.

Chapter 5 develops a distributed optimal feedback controller for minimizing the total acoustic energy in an enclosure. Our approach is to model the residual field as the sum of the outputs of two infinite-dimensional linear systems. The unwanted noise field is modeled as the output of an infinite-dimensional linear system driven by a white process. Similarly, the canceling field is modeled as the output of another infinite-dimensional linear system driven by the control signal. Finally, the residual field is modeled as the sum of the outputs of these two linear systems. The inputs to the controller are the outputs of ordinary microphones, and the control signal drives an ordinary loudspeaker. We show that the control signal that minimizes a certain class of performance criteria is a linear feedback of the estimated states of these two linear systems. These state estimates can be computed using a Kalman filter, and the feedback gain operator can be computed by iterating an operator Riccati equation. This is the first mathematically rigorous formulation of the distributed ANC problem known to us.

Lastly, in Chapter 6 we summarize the major contributions of this thesis. We also suggest several potentially important directions for future research.

Chapter 2

Previous Work on Active Noise Cancellation

2.1 Introduction

This chapter motivates our study of active noise cancellation and presents a summary of the previous work done in this area. We try to distinguish between the acoustic objectives of different ANC systems and the control strategies used to achieve these objectives. To appreciate the advantages and the limitations of active noise cancellation, it is necessary to understand the relevant acoustical principles as well as the relevant control strategies. Hence, this chapter includes a Section on acoustical principles behind ANC and a Section on control strategies used for ANC. The relationship between active noise cancellation and signal estimation is also studied in this chapter.

This chapter is organized as follows. Section 2.2 describes the physical basis for active noise cancellation, with emphasis on three dimensional sound fields. The different sound fields that are generated by ANC systems designed to achieve different acoustic objectives are studied in this Section using Modal Analysis techniques. In Section 2.3, we review several commonly used control strategies for ANC. Many examples of existing ANC systems are presented to illustrate the use of these control strategies. In Section 2.4 , the performance of two specific control strategies for ANC is analyzed in detail. Finally in Section 2.5, the

18

relationship between active noise cancellation and signal estimation is explored. A brief summary of this chapter is presented in Section 2.6.

2.2 Acoustical Principles

The fundamental problem in active noise cancellation is to generate an acoustic field that interferes destructively with the undesired noise [10, 13, 23, 40, 50, 59]. The undesired field is usually referred to as the "primary" field, and the interfering field is referred to as the "secondary" field. The question therefore arises as to whether two different sources of sound can generate the same acoustic field? If the answer to this question is yes and one source is under our control, a simple change of sign of the controlled source will cancel the primary field everywhere.

We shall give a simple example of two different sources that generate the same acoustic field. Let $q(\underline{x}, t)$ denote a sound source generating the pressure field $p(\underline{x}, t)$, where $\underline{x}$ is the spatial variable; t is the temporal variable; and the underline denotes a vector-valued quantity. Furthermore, assume that the source is only non-zero over some bounded spatial domain Γ. In this case, the wave equation for pressure is

$$\frac{1}{c^2} \frac{\partial^2 p(\underline{x}, t)}{\partial t^2} - \nabla^2 p(\underline{x}, t) = q(\underline{x}, t), \tag{2.1}$$

where ∇^2 is the three dimensional Laplacian, and c is speed of sound. The pressure field $p(\underline{x}, t)$ outside the region of support of the source will remain unchanged if the source field is supplemented by $\frac{1}{c^2} \frac{\partial^2}{\partial t^2} q(\underline{x}, t) - \nabla^2 q(\underline{x}, t)$. To see this, note that the resulting pressure field in this case, $p'(\underline{x}, t)$, must satisfy

$$\frac{1}{c^2} \frac{\partial^2}{\partial t^2} p'(\underline{x}, t) - \nabla^2 p'(\underline{x}, t) = q(\underline{x}, t) + \frac{1}{c^2} \frac{\partial^2}{\partial t^2} q(\underline{x}, t) - \nabla^2 q(\underline{x}, t), \tag{2.2}$$

and eq. (2.2) can be rewritten as

$$\frac{1}{c^2} \frac{\partial^2}{\partial t^2} (p'(\underline{x}, t) - q(\underline{x}, t)) - \nabla^2 (p'(\underline{x}, t) - q(\underline{x}, t)) = q(\underline{x}, t). \tag{2.3}$$

Comparing (2.3) to (2.1) and recalling that $q(\underline{x}, t)$ is zero outside Γ, we conclude that $p'(\underline{x}, t)$ and $p(\underline{x}, t)$ are equal outside Γ. Therefore, we have found an example of two distinct sources that produce the same acoustic pressure field.

19

More generally, Kempton [31] has shown that the sound field generated by any source can be reproduced by an appropriate infinite series of source singularities positioned at any desired point. Hence, complete cancellation of the primary field can be achieved by arranging these source singularities to produce the negative of the primary field.

A more obvious way to cancel the effect of an acoustic source is to use Kirchhoff's theorem. This theorem provides a formula by which the entire effect of sources inside a closed boundary can be duplicated by sources on this boundary. Hence, if the negative of these sources is used on the boundary, the primary field in the exterior of the boundary is completely canceled. Jessel and Angevine [28, 29] have proposed such an ANC system in theory. Unfortunately, the secondary source in this case is distributed, and a discrete approximation to this distributed source has not been developed yet. Inspired by this theory, several researchers have built experimental ANC systems with limited success [4, 32].

With active noise cancellation, the combined radiated output of the primary and secondary sources is typically much less than the radiated output of the primary source alone. The reduction in the combined radiated power is achieved by a reduction in the radiation impedance seen by the primary source and/or absorption of energy by the secondary source [17, 46].

Because of the practical difficulties associated with cancellation of an entire noise field, most researchers have focused on developing ANC systems with very limited acoustic objectives. The most commonly used acoustic objective is attenuation of pressure at a finite number of spatial points.

Modal analysis is one of the very few techniques avaiable for studying the global effects of ANC in an enclosed volume.

Modal Analysis for ANC

Modal analysis is a practical approach for analyzing the sound field generated by an acoustic system operating in an enclosure. The starting point is to assume that the sound field in the enclosure has a periodic time dependence $e^{j\Omega t}$. Let $P(\underline{x}, t)$ refer to the pressure at location

$\underline{x}$ and at time t. In this case, $P(\underline{x}, t)$ can be expressed as

$$P(\underline{x}, t) = p(\underline{x}, \Omega)e^{j\Omega t}, \tag{2.4}$$

where $p(\underline{x}, \Omega)$ is not a function of time. It is further assumed that $p(\underline{x}, \Omega)$ can be expressed in terms of a finite number of the normal modes of the enclosure

$$p(\underline{x}, \Omega) = \sum_{n=0}^{N} \Psi_n(\underline{x})a_n(\Omega), \tag{2.5}$$

where $\Psi_n(\underline{x})$ is the the n-th normal mode of the enclosure, and $a_n(\Omega)$ is the complex amplitude of the n-th mode.

Next, it is assumed that the sound field is linear so that the complex amplitude $a_n(\Omega)$ can be expressed as the sum of the contributions from the primary source and M secondary sources, i.e.

$$a_n(\Omega) = a_n^p(\Omega) + \sum_{m=1}^{M} B_{nm}(\Omega)q_m^s(\Omega), \tag{2.6}$$

where $a_n^p(\Omega)$ is the n-th modal amplitude produced by the primary source; $q_m^s(\Omega)$ is the net strength of the m-th secondary source; and $B_{nm}(\Omega)$ specifies the degree to which the m-th secondary source contributes to the n-th mode.

Equation (2.6) can be rewritten in vector form as

$$\underline{a} = \underline{a}^p + B\underline{q}^s, \tag{2.7}$$

where $\underline{a}$ is the $N \times 1$ vector

$$\underline{a} = \begin{bmatrix} a_1(\Omega) \\ a_2(\Omega) \\ \vdots \\ a_N(\Omega) \end{bmatrix}; \tag{2.8}$$

21

$\underline{a}^p$ is the $N \times 1$ vector

$$\underline{a}^p = \begin{bmatrix} a_1^p(\Omega) \\ a_2^p(\Omega) \\ \vdots \\ a_N^p(\Omega) \end{bmatrix} ; \tag{2.9}$$

$\underline{q}^s$ is the $M \times 1$ vector

$$\underline{q}^s = \begin{bmatrix} q_1^s(\Omega) \\ q_2^s(\Omega) \\ \vdots \\ q_M^s(\Omega) \end{bmatrix} ; \tag{2.10}$$

and B is the $N \times M$ matrix of complex modal coupling coefficients

$$B = \begin{bmatrix} B_{11}(\Omega) & \dots & B_{1M}(\Omega) \\ \vdots & \vdots & \vdots \\ B_{N1}(\Omega) & \dots & B_{NM}(\Omega) \end{bmatrix} . \tag{2.11}$$

The total time-averaged acoustic potential energy in the enclosure is denoted by E_p and is given by

$$E_p = \frac{1}{4c^2\rho} \int_V |p(\underline{x}, \Omega)|^2 d\underline{x}, \tag{2.12}$$

where c is the speed of sound, and ρ is the density of fluid in the enclosure. Recalling the orthonormality of Ψ_n's, we see that

$$E_p = \frac{1}{4c^2\rho} \sum_{n=1}^{N} |a_n(\Omega)|^2 \tag{2.13}$$

$$= \frac{1}{4c^2\rho} \underline{a}^H \underline{a}. \tag{2.14}$$

Using the above formulation, one can find the secondary source strength $\underline{q}^s$ that minimizes the acoustic potential energy E_p. Since E_p is a quadratic function of the source

22

strength vector $\underline{q}^s$, the following source strength vector minimizes the total acoustic potential energy in the enclosure

$$\underline{q}^s = -[B^H B]^{-1} B^H \underline{a}^p, \tag{2.15}$$

and the resulting minimum value of the potential energy is

$$E_p = \frac{1}{4c^2\rho}[(\underline{a}^p)^H \underline{a}^p - (\underline{a}^p)^H B[B^H B]^{-1} B^H \underline{a}^p]. \tag{2.16}$$

A similar formulation can be used to express the total acoustic kinetic energy in the enclosure, E_k, as a quadratic function of the source strength vector $\underline{q}^s$. Furthermore, one can find the secondary source strength that minimizes E_k or $(E_k + E_p)$. Note that $(E_k + E_p)$ is the total acoustic energy in the enclosure.

Nelson et al. [36] considered the problem of minimizing the acoustic potential energy in a rectangular enclosure. The three dimensional sound field in the enclosure was expressed as the sum of the contributions of 7000 modes. Given the position and the strength of a pure tone primary source (a point source), the authors used equation (2.15) to find the amplitude and phase of a number of secondary point sources that would minimize the total acoustic potential energy in the enclosure.

At frequencies above the Schroeder cut-off frequency of the enclosure, the authors found that substantial reductions in acoustic potential energy are not obtainable unless the canceling sources are separated from the primary source by a distance of no more than half the wave length [9].

At frequencies below the Schroeder cut-off frequency of the enclosure, the system exhibited several interesting features [19]. The authors found that appreciable reductions in the overall potential energy can be achieved by introduction of a small number of secondary sources spaced greater than half a wavelength from the primary source, provided that the enclosure is being excited at or near to acoustic resonance (where a single mode dominates the response). In the low frequency case, the location of the canceling sources was found to have a great influence on the performance of the system. It was found that at excitation frequencies where only a single mode dominates the response of the enclosure, very large reductions in potential energy can be achieved. On the other hand, at excitation frequencies

23

where many modes contribute to the response of the enclosure, a few secondary sources are unable to control these modes without increasing the excitation of a number of other modes; hence, little reduction in the total acoustic potential energy is achieved.

David [15] used a secondary point source to drive the acoustic pressure to zero at a single point in a rectangular enclosure and studied the result of this action on the rest of the sound field in the enclosure. He found that a "zone of quiet" around the cancellation point, within which the sound pressure level is reduced by more than 15dB, has a diameter of about one tenth of the wavelength of the excitation frequency. The mean-square pressure away from the point of cancellation was found to increase if the transfer impedance between the secondary source and the cancellation point was very small at the excitation frequency. This led the authors to conclude that the secondary sources should be placed close to the cancellation point to avoid increasing the mean-square pressure away from the cancellation point.

Similar studies have been performed for the free field case [37]. In these studies, the primary source and the secondary sources are assumed to be in a free field environment with the primary source radiating at a single known frequency. The secondary source strengths that minimize the power output of the combination of primary and secondary sources is then found, and the minimum value of the power output is calculated. It is found that significant reduction in the combined power output may only be achieved if secondary sources are placed within a distance of half a wavelength of the primary source.

It is important to note that in all these experiments the secondary source strengths were calculated based on the exact knowledge of the strength of the primary source for all time. Therefore, these results can only be used as rough guidelines for predicting the global behavior of practical ANC systems in which the source strength is unknown. In a practical ANC system, the source strength must be estimated causally based on the available measurements.

A typical ANC system utilizes several microphones to monitor the attenuated field and/or the primary field. The outputs of these microphones are used as inputs to some sort of electronic controller. The controller is designed to drive the secondary sources in

24

such a way that the desired acoustic objective is achieved. The desired acoustic objective of almost all existing ANC systems is to attenuate the sound pressure at a finite number of spatial points. In the next Section, we will concentrate on the algorithms that are used by the controller to achieve this specific acoustic objective.

2.3 Control Strategies for ANC

A recent bibliography of references for ANC contains over 3450 entries [27]; therefore, it is not possible to cover each reference individually. However, most of these references describe particular applications of ANC rather than new concepts. In this Section we concentrate on those control strategies that are most commonly used in active noise cancellation, and selected references are only cited to illustrate the use of these control strategies.

Most existing active noise cancellation controllers can be categorized as pointwise feedback controllers or as pointwise feedforward controllers. The specific pointwise feedback controller and pointwise feedforward controller that we focus on in this thesis are presented next.

2.3.1 Feedback Controllers for Active Noise Cancellation

A generic feedback controller of the type we focus on in this thesis is depicted in Fig. 2.1, where the input to the controller $e(t)$ is the sum of the plant output $c(t)$ and the stochastic disturbance $n(t)$. The goal is to choose the control signal $r(t)$, based on observations of $\{e(\tau) : \tau < t\}$, so that the residual signal at the plant output $e(t)$ follows a desired trajectory. Note that $n(t)$, $e(t)$ or $r(t)$ can be scalar valued or vector valued. The important assumption here is that the control signal is generated based on the measurements of the residual signal at the plant output, and not based on direct measurements of the disturbance. In this case, all the inputs to the controller contain a part due to the plant output.

An illustrative example of a feedback ANC system is the noise canceling headphone developed by Wheeler [51]. Wheeler used an analog feedback system to reduce the sound pressure fluctuations in a headphone close to the ear of the listener. The noise cancellation environment and the corresponding block diagram are depicted in Fig. 2.2 and Fig. 2.3,

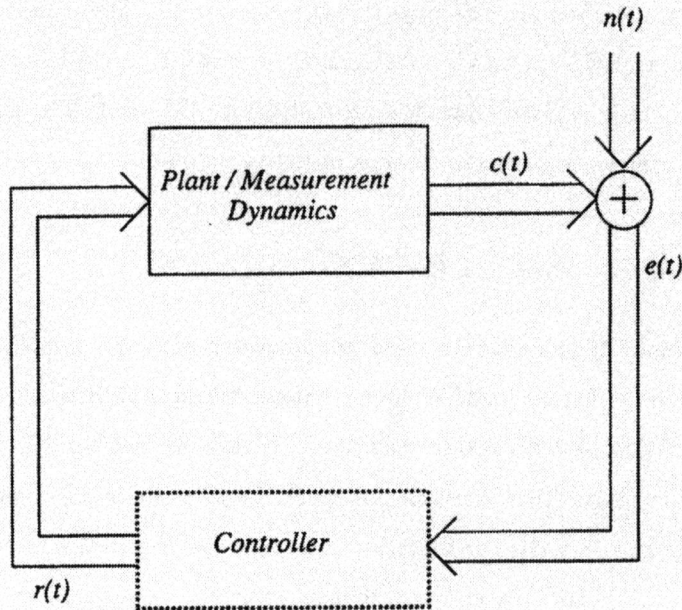

Figure 2.1: Generic feedback controller.

respectively.

Throughout this chapter we rely on the following model of the microphone. The output of the microphone is assumed to be the value of the acoustic pressure field at the location of the microphone, i.e. the microphone is assumed to measure the acoustic pressure at the location of the microphone without any distortion [6].

In Fig. 2.3, plant $G(s)$ represents the overall transfer function from the canceling loudspeaker input $r(t)$ to the value of the canceling field at the location of the microphone $c(t)$ and incorporates the transfer functions of the loudspeaker and the propagation path between the loudspeaker and the microphone. The microphone output $e(t)$ is the sum of the unwanted noise at the location of the microphone $n(t)$ and the canceling signal at the location of the microphone $c(t)$. The goal of the controller is to minimize the sound pressure fluctuations as measured by the microphone, i.e. the controller is designed to minimize $E\{e^2(t)\}$. Comparing Fig. 2.3 with Fig. 2.1, we see that the noise canceling headphone developed by Wheeler is a scalar analog feedback controller.

Note that this system attenuates the noise power at the microphone, and it is assumed

26

that attenuating the noise at the microphone will result in noise attenuation at the ear of the listener. In noise canceling headphones with low frequency noise, this assumption is quite reasonable because the ear of the listener is only about 3 centimeters form the microphone [15].

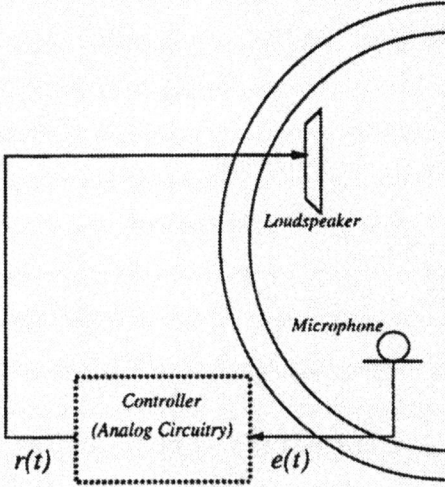

Figure 2.2: Noise canceling headphone proposed by Wheeler.

In the system of Fig. 2.3, the noise field at the microphone without the noise cancellation system is $n(t)$. With the noise cancellation system, the noise field at the microphone is $e(t)$. The power spectra of these two signals are related by the closed-loop transfer function $T(s)$ that is defined as

$$T(s) = \frac{1}{1 - H(s)G(s)} , \qquad (2.17)$$

where $H(s)$ is the transfer function of the controller. Specifically,

$$P_{ee}(j\Omega) = |T(j\Omega)|^2 P_{nn}(j\Omega), \qquad (2.18)$$

where $P_{ee}(e^{j\Omega})$ is the power spectrum of $e(t)$, and $P_{nn}(j\Omega)$ is the power spectrum of $n(t)$ with Ω being the continuous-time frequency variable. Looking at eq. (2.18), we see that to achieve attenuation over a certain frequency band, $|T(j\Omega)|$ must be made small over this

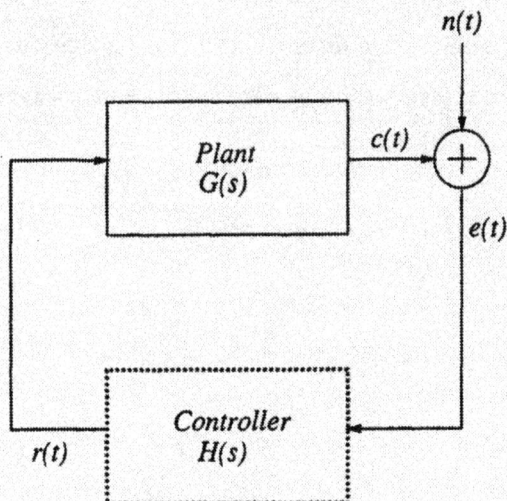

Figure 2.3: Block diagram of the noise canceling headphone proposed by Wheeler.

band. Typically, the analog controller is designed so that the closed-loop system achieves moderate noise attenuation (15dB) over a relatively wide range of frequencies (50Hz-500Hz) [12, 51]. Note that due to its analog nature, the controller in Fig. 2.2 cannot be changed once it is built.

A Discrete-Time Feedback Controller

A discrete-time version of the system in Fig. 2.2 has been proposed by Graupe [26]. Since the controller proposed by Graupe is a discrete-time controller we assume throughout this Section that the noise to be canceled, the canceling signal, and all other intermediate signals are appropriately sampled. Consequently in this Section, "t" represents the normalized sampling time. Note that up to this point, we have been using "t" as the continuous-time variable. Occasionally this notation can be ambiguous; however, we think that in most cases the reader will be able to resolve this ambiguity based on the context. Whenever this ambiguity cannot be resolved based on the context, we explicitly state whether "t" denotes the continuous-time variable or the discrete-time variable.

The block diagram for the system of Graupe is depicted in Fig. 2.4, where $G(z)$ is the overall transfer function from the canceling loudspeaker input $r(t)$ to the value of the

28

canceling field at the location of the microphone $c(t)$ and incorporates the transfer function of the loudspeaker and the propagation path between the loudspeaker and the microphone. The microphone output $e(t)$ is the sum of the unwanted noise $n(t)$ and the canceling signal $c(t)$. $n(t)$ is modeled as the output of the transfer function $\Phi(z)$ driven by the white noise process $w(t)$. The goal of the controller is to minimize the sound pressure at the microphone, i.e. the controller is designed to minimize $E\{e^2(t)\}$.

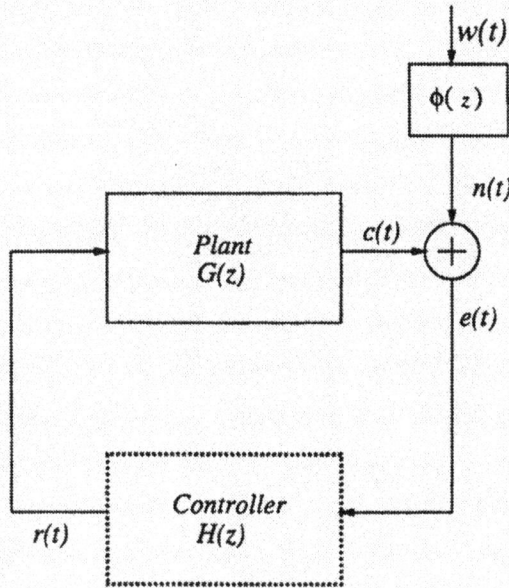

Figure 2.4: Block diagram of the ANC system proposed by Graupe.

A modified version of the Box-Jenkins controller [8] is used to continuously adjust $H(z)$ such that $E\{e(t)^2\}$ is minimized. This system uses the noise model $\Phi(z)$ to adapt the controller to the specific type of noise that is being canceled. Based on computer simulations, the authors claim that this system achieves 54dB of noise cancellation in the context of compressor noise.

Incorporation of a stochastic model for the unwanted noise in the system proposed by Graupe is one reason for the much better performance of this system than the performance of the system proposed by Wheeler. The digital system of Graupe uses the model of the unwanted noise to determine those frequencies at which the noise signal has significant energy, and it adjusts $H(z)$ such that the resulting closed-loop transfer function, $T(e^{jw}) = \frac{1}{1-H(e^{jw})\,G(e^{jw})}$, is small over these frequencies. Since the compressor noise is extremely narrow band, the resulting closed-loop system essentially acts as a notch filter. On the other hand, the analog system proposed by Wheeler is fixed and makes no use of the spectral characteristics of the unwanted noise; therefore, it achieves modest noise attenuation over a wide range of frequencies.

The second reason for the excellent results obtained by Graupe is that his simulations were performed with a $G(z)$ that did not contain any delay. Moreover, the controller proposed by Graupe is derived based on the explicit assumption that $G(z)$ contains no delay. This severely limits the applications in which this algorithm can be used, since there are many scenarios in which the propagation delay between the loudspeaker and the error microphone is of the order of few sample intervals.

2.3.2 Feedforward Controllers for Active Noise Cancellation

A generic feedforward controller of the type we focus on in this thesis is depicted in Fig. 2.5, where $q(t)$ is the input of the controller, and $n(t)$ is the stochastic disturbance. The goal is to choose the controller such that the residual signal at the plant output $e(t)$ is as close as possible to a desired trajectory. In this case, the input of the controller contains no part due to the plant output $c(t)$, i.e. there is no feedback from the plant output to the inputs of the controller. This is in contrast to the feedback scheme of the previous Section where the input of the controller was the sum of the plant output and the disturbance.

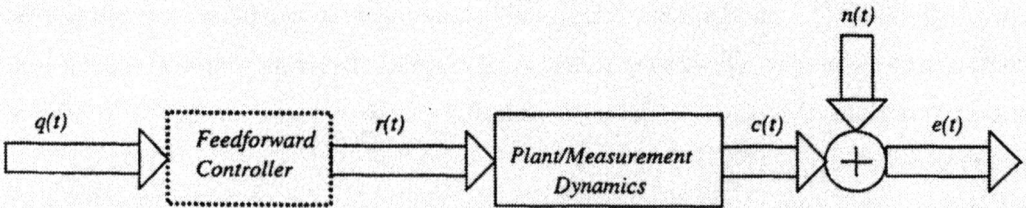

Figure 2.5: Generic feedforward controller.

Single-Channel Feedforward ANC Systems

Most of the early work in ANC was related to attenuation of duct noise at low frequencies using feedforward controllers. A very thorough analysis of the duct problem was carried out by Swinbanks in 1973 [49]. We shall present his work in some detail next because it illustrates many of the issues involved in using feedforward controllers for ANC. The system developed by Swinbanks is a continuous-time system; therefore, all the signals involved in this Section are continuous-time signals.

Figure 2.6 is a simplified diagram of the system proposed by Swinbanks. The primary wave, propagating in the down stream direction of an infinitely long duct with square cross Section, is detected by a microphone which supplies the information necessary for the operation of the controller. The controller generates a secondary field in zone 3 which is the negative of the primary field in this zone. For the microphone to only detect the primary field, the secondary field must not propagate upstream towards zone 1. Moreover, if there were any upstream obstructions in zone 1, the secondary field propagating upstream would get reflected back to zone 3, and one would have to deal with the problem of controlling both the primary wave and the reflected wave in zone 3.

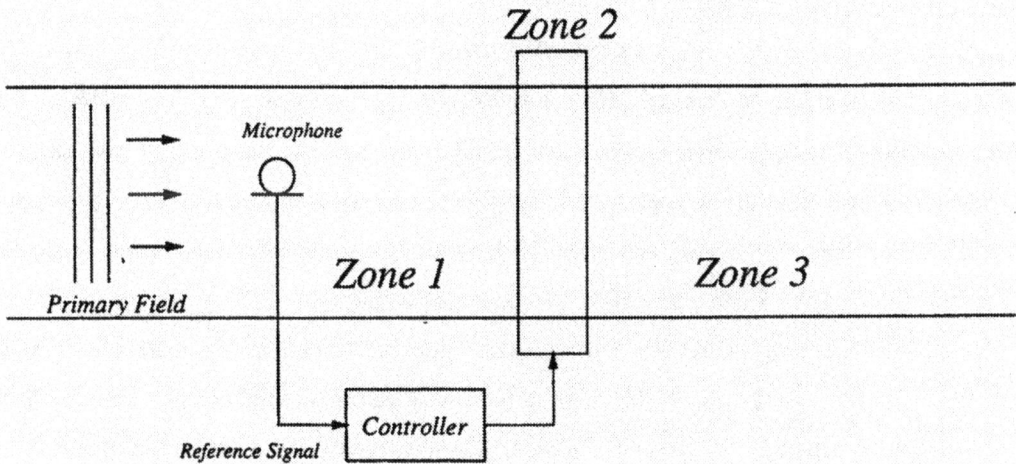

Figure 2.6: Sketch of the ANC system proposed by Swinbanks.

Swinbanks showed that by using a single ring sources, consisting of four point sources, it is possible to generate an output consisting only of a propagating plane wave for frequencies up to 2.1 times the cut-off frequency of the duct, see Fig. 2.7. More precisely, for these frequencies, the non-plane wave modes in the duct will decay exponentially along the long axis of the duct away from the ring source. Swinbanks also showed that by combining the effects of two such ring sources, it is possible to generate an approximately unidirectional plane wave in the duct up to 2.1 times the cut-off frequency of the duct, see Fig. 2.7. Specifically, if the source strength of the upstream ring is $m_1(t)$ and the strength of the downstream ring is $m_2(t)$, there will be approximately zero output in the upstream direction provided that

$$m_1(t) = -m_2(t - \tau_0), \tag{2.19}$$

where τ_0 is an appropriate amount of delay which depends on the separation between the two rings and the speed of sound in the duct.

As illustrated in Fig. 2.7, the controller in this case consists of a delay element $e^{-s\tau_1}$ followed by an analog filter whose purpose is to equalize the combined frequency response of the secondary sources. Let $p(x,t)$ denote the pressure field at location x (measured along the long axis of the duct) and at time t. Over the frequency range of interest, the equalizer tries to make $p(b,t) = m_2(t - \tau_2)$ for some delay τ_2.

The operation of the ANC system of Swinbanks can be summarized as follows. At time t, the controller predicts $p(b, t + \tau_2)$ and drives the equalized secondary sources with the negative of this predicted value. If the prediction is perfect, the primary field and the secondary field will interfere destructively at $x = b$, and the result will be complete silence at this point. However, every other point in zone 3 will also be completely silent, since the residual field in zone 3 is a plane wave propagating to the right.

Implementation of Swinbanks method, with minor modifications, resulted in a flurry of publications (e.g., [7, 11, 41, 44]). These implementations showed that the main drawback of Swinbanks' method is that very precise analog electronics are needed to implement the feedforward controller. The required degree of precision in the magnitude and the phase of the controller was found to be impossible to maintain except over a very limited range of frequencies. In short, Swinbanks' method was found to lack robustness.

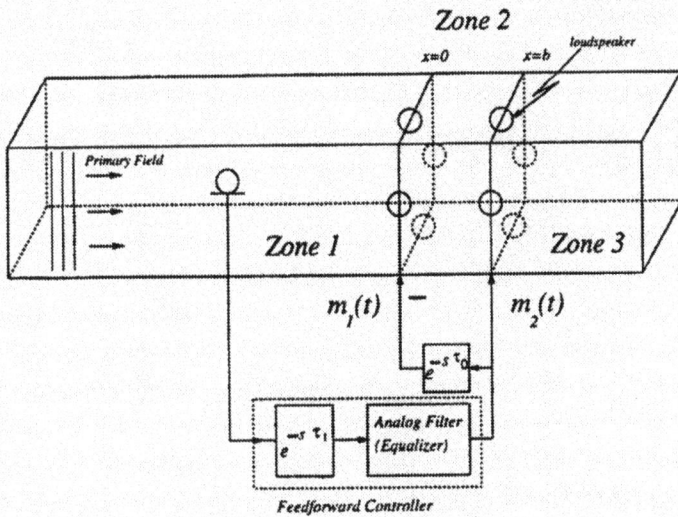

Figure 2.7: Block diagram of the ANC system proposed by Swinbanks.

The most interesting feature of the system proposed by Swinbanks is that it achieves complete noise cancellation at *every* point in the downstream Section of the duct, although the controller is designed to cancel the noise at a single point. This is a direct consequence of the unidirectionality of the primary and secondary fields. In most applications of ANC, the sound fields involved are diffused; hence, the method of Swinbanks often cannot be used.

In 1981 Burgess proposed an adaptive discrete-time feedforward controller for ANC [10]. The system proposed by Burgess is quite robust and does not rely on the unidirectionality of the sound fields involved. Therefore, this system can be used in a wide range of applications. The acoustic goal of this system is to attenuate the noise at the location of a microphone based on the information provided by a reference signal.

As depicted in Fig. 2.8, the system proposed by Burgess consists of an error microphone and a canceling loudspeaker. The goal is to to *attenuate* the noise at the location of the error microphone. The reference signal $q(t)$ provides information about the unwanted noise at the error microphone and is the input of the controller H. If the noise is generated by a pump or an engine, the reference signal is typically the output of a non-acoustical sensor attached

33

to the pump or the engine (e.g., a tachometer on the engine fly wheel). The controller is a discrete-time finite impulse response system whose coefficients are continuously adjusted to minimize the power of the residual signal $e(t)$ at the error microphone.

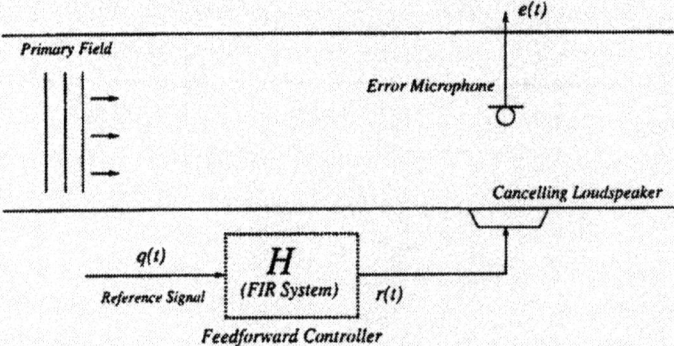

Figure 2.8: ANC system proposed by Burgess.

The block diagram for the system in Fig. 2.8 is depicted in Fig. 2.9, where $q(t)$ is the reference signal; H is the FIR controller whose coefficients at time "t" are $\{h(i,t)\}_{i=1}^{L}$; and $G(z)$ is the transfer function from the input of the loudspeaker $r(t)$ to the value of the canceling field at the location of the error microphone $c(t)$. The output of the error microphone $e(t)$ is the sum of the unwanted noise at the location of the microphone $n(t)$ and the value of the canceling field at the location of the microphone $c(t)$. This block diagram is derived under the assumption that there is no feedback from the output of the canceling loudspeaker to the reference signal.

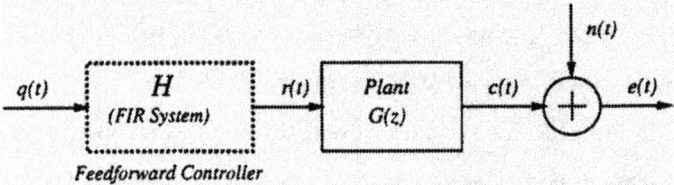

Figure 2.9: Block diagram of the ANC system proposed by Burgess.

Assuming that $G(z)$ is known, an LMS-type algorithm [52] can be derived for adjusting the coefficients $\{h(i,t)\}_{i=1}^{L}$ such that the mean-square power of the residual signal $e(t)$ is

34

minimized. The update equations for the coefficients of the controller are:

$$h(i, t+1) = h(i,t) - 2\,\mu\,e(t)\,v(t-i) \qquad i = 1,\dots,L \quad , \tag{2.20}$$

where $v(t)$ is defined as $v(t) \stackrel{\text{def}}{=} q(t) * g(t)$, and μ is the step-size of the algorithm. Note that $v(t)$ can be computed by "filtering" the known signal $q(t)$ with the known impulse response $g(t)$; hence, this algorithm is sometimes referred to as the "filtered-x" algorithm [53].

Multiple-Channel Feedforward ANC Systems

A multiple-channel version of the algorithm of Burgess has been developed by Elliot *et. al* [21, 18]. This algorithm is a discrete-time algorithm; therefore, all signals involved are assumed to be appropriately sampled. As depicted in Fig. 2.10, this system consists of L microphones and M loudspeakers. The controller in this case is a bank of M finite impulse response filters $\{H_i\}_{i=1}^{M}$ where each filter is driven by the reference signal $q(t)$. The goal is to adjust the coefficients of these filters such that a cost function involving the mean-square sum of the microphone outputs and the mean-square sum of the control signals is minimized. Specifically, given the matrix of transfer functions from the loudspeaker inputs to the microphone outputs, the authors develop an LMS-type algorithm for adjusting the coefficients of these M filters such that

$$E\left\{\sum_{i=1}^{L} e_i^2(t) + \alpha \sum_{j=1}^{M} r_j^2(t)\right\}$$

is minimized, where α is a positive weighting coefficient for the control effort. Note that the cost function used here is a bit more general than the one used by Burgess [10].

The influence of the loudspeaker transfer function and acoustic delay on the performance of this algorithm has been thoroughly analyzed in a number of papers (e.g., [20, 47, 48]).

2.4   Performance Analysis for Two ANC Controllers

The performance of the continuous-time single-channel feedback ANC system and the performance of the continuous-time single-channel feedforward ANC system are analyzed in this Section.

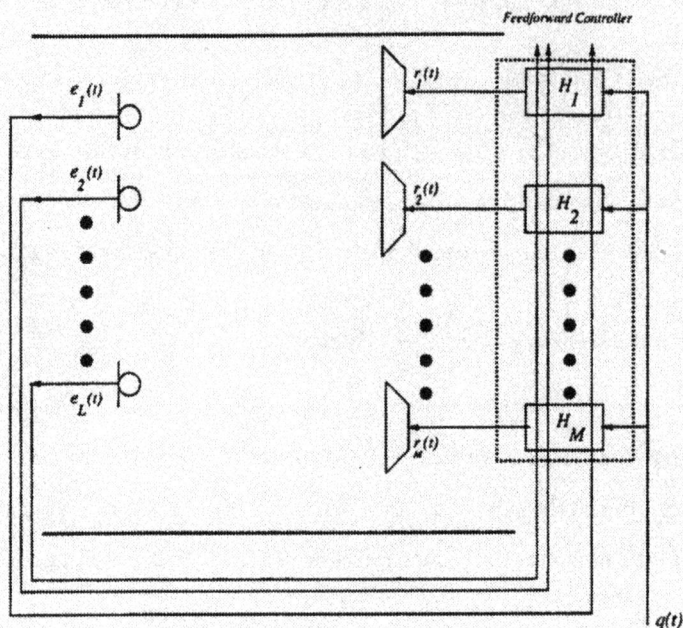

Figure 2.10: Multiple-channel feedforward ANC system.

2.4.1 Optimal Single-Channel Feedback Controller

Further insight about the continuous-time feedback ANC system of Wheeler [51] can be gained by finding the optimal feedback controller in this case. Specifically in Fig. 2.3, we would like to find the causal controller $H(s)$ that minimizes $E\{e^2(t)\}$, assuming that the noise $n(t)$ is stationary with known power spectrum $P_{nn}(j\Omega)$ and assuming that the plant $G(s)$ is known. Our strategy is to first find the minimizing controller within a restricted class of controllers, and then show that no other controller can outperform this controller. We shall make the further assumption that $G(s)$ is stable. This is a reasonable assumption, since the physical system that $G(s)$ corresponds to is passive.

Let us start by looking at controllers of the form depicted in Fig. 2.11. In this figure, $G(s)$ (the transfer function of the plant) is fixed, and $Q(s)$ is a causal and stable system that can be varied subject to the restriction that the resulting closed-loop system must be stable. Let us take an arbitrary controller within this class and assume that this controller results

36

from some $Q(s)$, then the closed-loop transfer function corresponding to this controller is

$$T(s) \;=\; \frac{1}{1 - G(s)\,H(s)} \tag{2.21}$$

$$=\; \frac{1}{1 - G(s)\left(\frac{Q(s)}{1 + G(s)\,Q(s)}\right)} \tag{2.22}$$

$$=\; 1 + G(s)Q(s). \tag{2.23}$$

It is important to note that since both $Q(s)$ and $G(s)$ are stable, there is no unstable pole/zero cancellation in the closed-loop system.

With this controller, the power spectrum of the residual signal $e(t)$ can be expressed in terms of the power spectrum of the unwanted noise $n(t)$ as

$$P_{ee}(j\Omega) = |1 + G(j\Omega)\,Q(j\Omega)|^2\, P_{nn}(j\Omega). \tag{2.24}$$

Therefore, the first step is to find the causal and stable $Q(s)$ in eq. (2.24) that minimizes the power of the residual signal $e(t)$ for a given $P_{nn}(s)$.

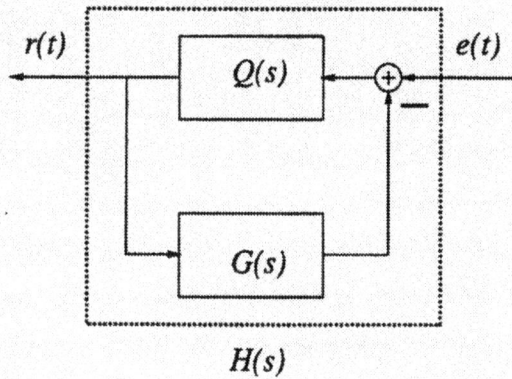

Figure 2.11: Restricted class of controllers.

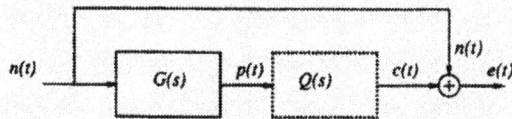

Figure 2.12: Equivalent block diagram of the the closed-loop system.

A convenient way to find the minimizing choice for $Q(s)$ is to represent equation (2.23) in block diagram form as depicted in Fig. 2.12. Looking at Fig. 2.12, it is clear that the choice for $Q(s)$ that minimizes $E\{e^2(t)\}$ is the causal Wiener filter [11] that produces the linear least-squares estimate (LLSE) of $-n(t)$ based on $\{p(\tau) : \tau \leq t\}$. Let us denote this particular choice for $Q(s)$ by $Q_{win}(s)$, the corresponding controller by $H_{win}(s) = \frac{Q_{win}(s)}{1+G(s)Q_{win}(s)}$, and the resulting $E\{e^2(t)\}$ by E_{win}. From eq. (2.23), we see that the closed-loop system in this case is also stable, since $G(s)$ and $Q_{win}(s)$ are both stable.

It remains to be shown that no other causal controller can outperform $H_{win}(s)$. We will prove this by contradiction. To this end, assume that such a controller $H^{\#}(s)$ exists, and the resulting value of $E\{e^2(t)\}$ is $E^{\#}$, which is less than E_{win}. It can be easily shown that this leads to a contradiction by choosing

$$Q(s) = \frac{H^{\#}(s)}{1 - H^{\#}(s)G(s)} \tag{2.25}$$

in Fig. 2.12 and noting that the resulting value for $E\{e^2(t)\}$ is $E^{\#}$, which is assumed to be less than E_{win}. This is a contradiction, since E_{win} is obtained by minimizing $E\{e^2(t)\}$ over all possible causal choices for $Q(s)$. Note that $Q(s)$ in eq. (2.25) is causal, since it can be realized as feedback interconnection of $H^{\#}(s)$ and $G(s)$. Using the uniqueness of the linear least-squares estimate, it can be shown that the canceling signal $c(t)$ that minimizes $E\{e^2(t)\}$ is unique.

This analysis shows that with the optimal controller, the canceling signal $c(t)$ equals the LLSE of $-n(t)$ based on $\{p(\tau) : \tau \leq t\}$, where $p(t) = n(t) * g(t)$. This analysis can be taken one step further. To this end, let us express $G(s)$ as

$$G(s) = G_{all}(s)G_{min}(s), \tag{2.26}$$

where $G_{all}(s)$ is an all-pass system function, and $G_{min}(s)$ is a minimum-phase system function. We also define $p'(t)$ as

$$p'(t) = n(t) * g_{all}(t). \tag{2.27}$$

The LLSE of $n(t)$ based on $\{p(\tau) : \tau \leq t\}$ is the same as the LLSE of $n(t)$ based on $\{p'(\tau) : \tau \leq t\}$, since $p(t) = p'(t) * g_{min}(t)$ and $G_{min}(s)$ is minimum-phase. Recalling that the minimizing choice for $c(t)$ is the LLSE of $-n(t)$ based on $\{p(\tau) : \tau \leq t\}$, we see that the minimizing choice for $c(t)$ can be alternatively expressed as the LLSE of $-n(t)$ based on

$\{p'(\tau) : \tau \le t\}$. Hence, the optimal performance of the noise cancellation system depends on the joint second-order statistics of $\{n(t), p'(t)\}$, which can be calculate form the power spectrum of $n(t)$, since $p'(t) = n(t) * g_{all}(t)$ and $g_{all}(t)$ is known. Clearly, noise attenuation will be high whenever $p'(t)$ is highly correlated with $n(t)$.

If the phase of $G_{all}(s)$ is approximately linear over the frequencies for which $n(t)$ has significant energy, the problem will reduce to predicting $-n(t)$ based on $\{n(\tau) : \tau \le t - \tau_0\}$, where τ_0 is the slope of the linear approximation to the phase of $G_{all}(s)$.

2.4.2   Optimal Single-Channel Feedforward Controller

Further insight about the feedforward controller of Burgess can be gained by looking at the system in continuous-time. The block diagram for the system in continuous-time is depicted in Fig. 2.13, where $q(t)$ is the reference signal; $H(s)$ is a continuous-time LTI controller; and $G(s)$ is the transfer function from the loudspeaker input $r(t)$ to the value of the canceling field at the location of the error microphone. The output of the error microphone $e(t)$ is the sum of the unwanted noise $n(t)$ and the canceling signal $c(t)$.

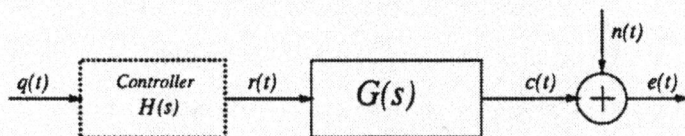

Figure 2.13: Block diagram of the feedforward ANC system in continuous time.

Let us express $G(s)$ as

$$G(s) = G_{all}(s)G_{min}(s), \tag{2.28}$$

where $G_{all}(s)$ is an all-pass system function and $G_{min}(s)$ is a minimum-phase system function. Assume that $n(t)$ and $q(t)$ are jointly stationary. In Fig. 2.13, $G(s)$ and $H(s)$ can be interchanged without affecting $c(t)$ or $e(t)$. The resulting system is depicted in Fig 2.14.

In Fig. 2.14, the choice for $H(s)$ that minimizes $E\{e^2(t)\}$ is the causal Wiener filter [57] that produces the linear least-squares estimate of $-n(t)$ based on $\{p(\tau) : \tau \le t\}$. The minimum value of $E\{e^2(t)\}$ can be calculated based on the joint second-order statistics of

39

$\{n(t), p(t)\}$, which in turn can be calculated based on the joint second-order statistics of $\{n(t), q(t)\}$, since $p(t) = q(t) * g(t)$ and $g(t)$ is known.

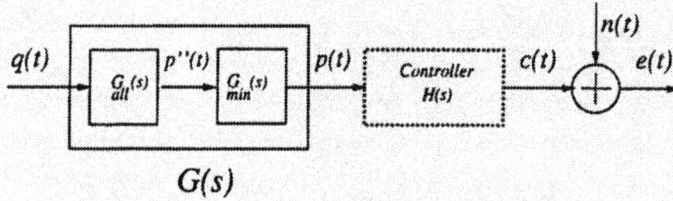

Figure 2.14: Alternate block diagram for the ANC system proposed by Burgess.

This analysis can be taken one step further by observing that the linear least-squares estimate of $n(t)$ based on $\{p(\tau) : \tau \leq t\}$ is equivalent to the linear least-squares estimate of $n(t)$ based on $\{p''(\tau) : \tau \leq t\}$, since $p(t) = p''(t) * g_{min}(s)$ and $G_{min}(s)$ is minimum phase. Therefore, the noise attenuation will be high whenever $p''(t)$ is highly correlated with $n(t)$.

If the phase of $G_{all}(s)$ is approximately linear over those frequencies for which $q(t)$ has significant energy, the problem will reduce to predicting $-n(t)$ based on $\{q(\tau) : \tau \leq t - \tau_0\}$, where τ_0 is the slope of the linear approximation to the phase of $G_{all}(s)$. In this case, the noise attenuation is substantial whenever $q(t)$ is highly correlated with the future values of $n(t)$. Note that this approximation is quite reasonable whenever the noise to be canceled is relatively narrow band.

A major shortcoming of the ANC system developed by Burgess is discussed next. The above analysis implies that the system of Burgess works well whenever the reference signal is highly correlated with the noise at the error microphone, and there is no feedback from the canceling loudspeaker to the reference signal. However, the formulation of Burgess does not tell us how to obtain such a reference signal. The author suggests the use of non-acoustical sensors attached to the noise source. In many applications, the source of the noise is inaccessible or unknown; hence, it is often not possible to acquire the reference signal in this way.

A common practice advocated by many researchers is to use the output of a second microphone (called the reference microphone) as the reference signal. Unfortunately, there is typically some feedback from the canceling loudspeaker to the output of the reference

40

microphone. In the presence of this feedback, the algorithm proposed by Burgess no longer minimizes $E\{e^2(t)\}$; moreover, the overall system has been observed to go unstable in the presence of this feedback. To decrease the feedback from the canceling loudspeaker to the reference microphone, the reference microphone is usually placed far away form the canceling loudspeaker and the error microphone. However, this decreases the correlation between the noise at the error microphone and the noise at the reference microphone, resulting in low levels of noise attenuation.

The multiple-channel feedforward ANC algorithm proposed by Elliot *et. al* [21, 18] suffers from the same shortcoming that the single-channel algorithm of Burgess does. Specifically, there is the problem of how to acquire a reference signal that is highly correlated with the unwanted noise and at the same time avoid acoustic feedback from the output of the canceling loudspeakers to the reference signal.

To circumvent the acoustic feedback problem associated with the use of a reference microphone, Eriksson *et. al.* [23] proposed an ANC system that takes the effect of the feedback into account in the design of the adaptive controller. Specifically, Eriksson proposed canceling the effect of the feedback at the output of the reference microphone, i.e. the input of the controller in this case is the sum of the output of the reference microphone and the negative of the estimate of the feedback. The resulting system uses two LMS algorithms and is quite complicated. More importantly, instability can result if the parameters of the adaptive algorithm are arbitrarily set, and the authors offer no procedure for properly setting these parameters. Furthermore, this system still works poorly whenever the noise at the reference microphone is not highly correlated with the noise at the error microphone.

2.5 Relationship between Signal Estimation and Active Noise Cancellation

The relationship between signal estimation and active noise cancellation is explored in this Section.

Signal estimation refers to estimating one stochastic process from observations of another

41

stochastic process. The goal in signal estimation is to generate, based on an observed signal, an output signal that is as close as possible to some desired signal. In a large class of signal estimation problems, the desired signal and the output signal are both numerical representation of some underlying physical signal. For example, in speech enhancement the desired signal is a numerical representation of the pressure fluctuations produced by the vocal cords of the speaker, and the output of the speech enhancement system is a numerical estimate of these pressure fluctuations. In this class of problems, all the signals involved (including the desired signal) are numerical; hence, there is no need to consider the physical signals that these numerical signals represent. Moreover in this class of problems, the process by which the output signal is obtained is purely numerical. In the reminder of this Section the term "signal estimation" refers exclusively to this type of numerical estimation.

In active noise cancellation, the desired signal is a physical signal and not a numerical signal; therefore, the output signal in active noise cancellation is also a physical signal. Furthermore, the output signal in ANC is generated as the sum of a canceling signal and one of the inputs. Since the output signal is a physical signal, the canceling signal and the input signal to which the canceling signal is added are both physical signals as well. Hence, every active noise cancellation system must have a transducer for physically generating the canceling signal. If the input of this transducer is significantly different from its output, the design of the ANC system must take the effect of the transducer into account.

For example, in acoustic active noise cancellation, the canceling signal and the unwanted noise are both physical signals corresponding to the primary pressure field and the secondary pressure field, respectively. The canceling signal is generated by a loudspeaker and is added physically to the unwanted noise through the interference of the primary field and the secondary field. Note that the process by which the output signal is obtained is *not* purely numerical, since the output signal is obtained through physical interference of two acoustic fields.

The relationship between signal estimation, in which the desired signal is a numerical signal, and active noise cancellation, in which the desired signal is a physical signal, is explored in this Section. In Section 2.5.1, a very general class of single-input/single-output signal estimation systems is compared to the class of single-channel feedback ANC systems. In Section 2.5.2, the class of single-channel feedforward ANC systems is compared to a special class of two-input/single-output signal estimation systems.

2.5.1 Relationship between Single-Channel Signal Estimation and Single-Channel ANC

The relationship between single-channel signal estimation and single-channel feedback active noise cancellation is studied in this Section.

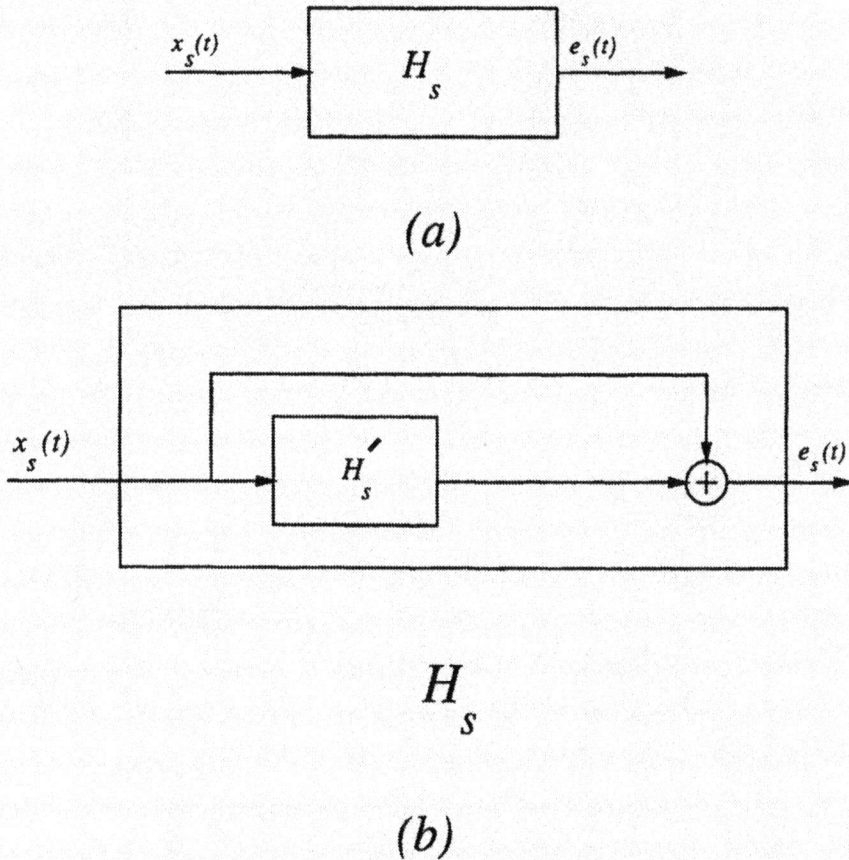

(a)

H_s

(b)

Figure 2.15: General single-channel signal estimation system.

Let us consider a completely general single-input/single-output signal estimation system. The estimator can be causal, non-causal, linear, or non-linear. As depicted in Fig. 2.15(a),

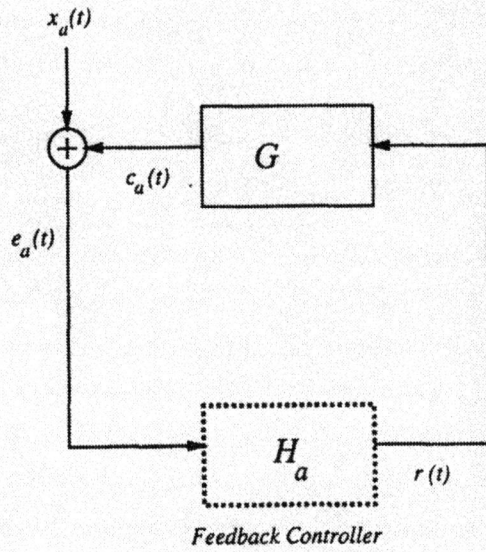

(a)

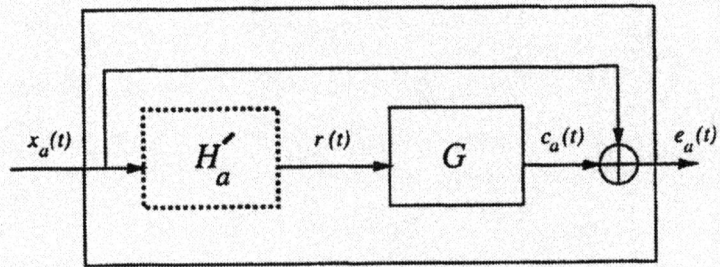

(b)

Figure 2.16: Feedback single-channel ANC system.

this system is a single-input/single-output system with input $x_s(t)$ and output $e_s(t)$. The objective is to find the system H_s that makes $e_s(t)$ as close as possible to some desired signal $d_s(t)$. Alternatively, the single-channel signal estimation system can be expressed in block diagram form as depicted in Fig. 2.15(b), where $H'_s = H_s - I$ with I being the identity system. Referring to Fig. 2.15(b), the single-channel signal estimation *problem* can be stated as follows:

Find the single-input/single-output system H'_s that minimizes

$$J\left(\{[x_s(\tau) + H'_s(x_s(\tau)), d_s(\tau)] : \ 0 \le \tau \le T\}\right), \tag{2.29}$$

where $J(.)$ is a pre-specified cost function that indicates how close the output $e_s(t)$ is to the desired signal $d_s(t)$ over the time interval $0 \le t \le T$, e.g.

$$J\left(\{[x_s(\tau) + H'_s(x_s(\tau)), d_a(\tau)] : 0 \le \tau \le T\}\right) = \tag{2.30}$$

$$\int_0^T E\left\{[x_s(\tau) + H'_s(x_s(\tau)) - d_s(\tau)]^2\right\} d\tau. \tag{2.31}$$

As depicted in Fig. 2.16(a), a single-channel feedback ANC system is a single-input/single-output feedback system with input $x_a(t)$ and output $e_a(t)$. The output $e_a(t)$ is generated as the sum of the canceling signal $c_a(t)$ and the input signal $x_a(t)$. Moreover, the canceling signal $c_a(t)$ is generated as the output of system G with input $r(t)$, and it is assumed that G is not an identity system. In this system, $x_a(t)$ and $c_a(t)$ are both physical signals, and system G represents the effect of the transducer that physically generates the canceling signal.

The objective of the single-channel feedback ANC system is to make $e_a(t)$ as close as possible to some desired signal $d_a(t)$. Specifically, the single-channel feedback ANC *problem* is to find the system H_a in Fig. 2.16(a) that makes $e_a(t)$ as close as possible to $d_a(t)$. Following the same steps used for the derivation of the optimal feedback ANC system in Section 2.4.1, it can be shown that the single-channel feedback ANC problem can be equivalently stated as finding the system H'_a in Fig. 2.16(b) that makes $e_a(t)$ as close as possible to the desired signal $d_a(t)$. Referring to Fig. 2.16(b), the single-channel ANC *problem* can be stated as:

Find the single-input/single-output system H_a' that minimizes

$$J\left(\left\{\,[x_a(\tau) + (GoH_a')(x_a(\tau)), d_a(\tau)] : \; 0 \leq \tau \leq T\right\}\right), \tag{2.32}$$

where "o" is the composition operator.

Comparing the single-channel signal estimation problem to the single-channel feedback ANC problem, we see that the ANC problem can be viewed as a constrained version of the signal estimation problem. Specifically, the single-channel feedback ANC problem can be formulated as finding H_s' that minimizes the expression in eq. (2.29) subject to the constraint that H_s' must be of the following form

$$H_s' = GoH_a', \tag{2.33}$$

for some system H_a' and a given system G.

We now compare the single-channel signal estimation system to the single-channel feedback ANC system in terms of their input/output characteristics. To this end, we assume that both systems are driven by the same input and that the desired signal for both systems is the same. We also assume that each system is designed so that the the distance between its output and the desired signal is minimized.

Under these assumptions, the performance of the ANC system is never better than the performance of the corresponding signal estimation system, since the optimal ANC system is obtained by minimizing $J(.)$ subject to a constraint, while the optimal signal estimation system is obtained by minimizing $J(.)$ without any constraints.

Again under these assumptions, if the minimizing H_s' in eq. (2.29) happens to equal (GoQ) for some system Q, the optimal ANC system will be identical to the optimal signal estimation system. Specifically, in this case, the minimizing H_a' in the ANC problem is system Q. For example, if G is invertible, the minimizing H_s' can be expressed as

$$H_s' = IoH_s' \tag{2.34}$$

$$= Go(G^{-1}oH_s'), \tag{2.35}$$

where I is the identity system. This implies that the solution to the single-channel feedback ANC problem can be expressed in terms of the solution to the corresponding single-channel signal estimation problem with $H_a' = G^{-1}oH_s'$.

The above discussion implies that unless the solution to the signal estimation problem is $(G o H'_a)$ with H'_a begin the solution to the corresponding ANC problem, the optimal feedback ANC system will be different from the optimal signal estimation system.

In general, a procedure for solving the single-channel signal estimation problem cannot be used to solve the single-channel feedback ANC problem, since a procedure for solving an un-constrained minimization problem cannot be used to solve a constrained minimization problem.

If the minimizing H'_a in the ANC problem can be commuted with G, the single-channel feedback ANC problem can be solved by solving a related two-input/single-output signal estimation problem. The ANC system in this case can be equivalently expressed in block diagram form as depicted in Fig. 2.17 where $q_a(t) = G(x_a(t)$. Referring to Fig. 2.17, the single-channel ANC problem in this case can be restated as finding the system H'_a that makes the output $e_a(t)$ as close as possible to the desired signal. This class of two-input/single-output signal estimation systems is studied in the next Section.

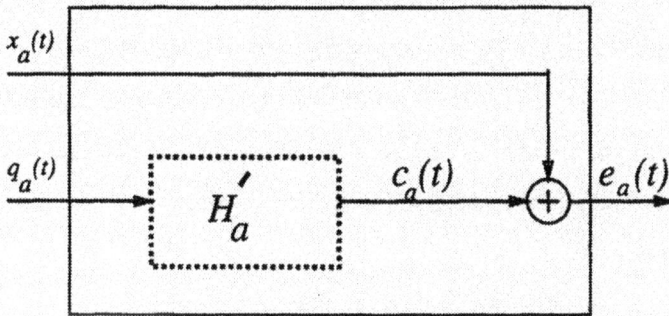

Figure 2.17: Equivalent block diagram for the single-channel feedback ANC system, provided that G and H'_a can be commuted.

2.5.2 Relationship between Feedforward Signal Estimation and Feedforward ANC

One way to do signal estimation is to first generate a canceling signal, and then form the estimate as the sum of this canceling signal and one of the observations. We shall refers to this approach as feedforward signal estimation.

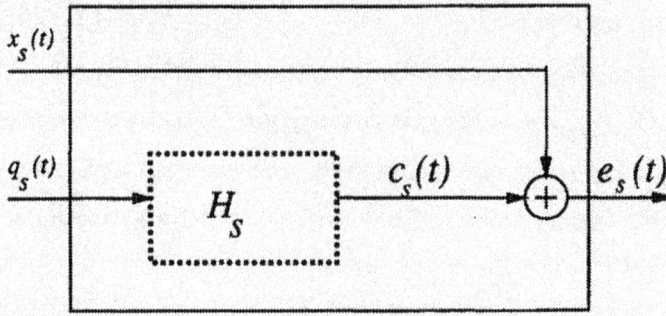

Figure 2.18: Feedforward signal estimation system.

A generic block diagram for a single-channel feedforward signal estimation system is depicted in Fig. 2.18. As depicted in Fig. 2.18, this system is a two-input/single-output system in which the output $e_s(t)$ is the sum of the canceling signal $c_s(t)$ and the input signal $x_s(t)$. Moreover, the canceling signal is generated solely based on one of the inputs, namely $q_s(t)$.

The adaptive noise canceling system proposed by B. Widrow [52] has the same structure as the system in Fig. 2.18, except that in Widrow's system H_s is continuously adjusted.

The objective of the feedforward signal estimation system is to make $e_s(t)$ as close as possible to some desired signal $d_s(t)$. Specifically, the feedforward signal estimation *problem* can be stated as follows:

Find the single-input/single-output system H_s that minimizes

$$J\left(\{[x_s(\tau) + H_s(q_s(\tau)), d_s(\tau)] : \ 0 \leq \tau \leq T\}\right). \tag{2.36}$$

Next, we consider the single-channel feedforward ANC system. As depicted in Fig. 2.19, a single-channel feedforward ANC system is a two-input/single-output system where the output $e_a(t)$ is generated as the sum of the canceling signal $c_a(t)$ and the input signal $x_a(t)$. Moreover, the canceling signal $c_a(t)$ is generated as the output of system G, and it is assumed that G is not an identity system. In this system, $x_a(t)$ and $c_a(t)$ are both physical signals, and system G represents the effect of the transducer that physically generates the canceling signal.

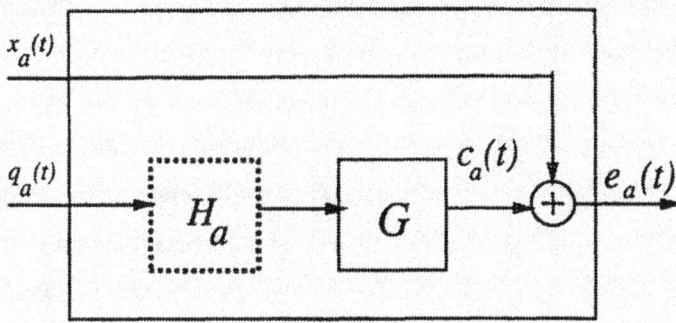

Figure 2.19: Feedforward ANC system.

The objective of the single-channel feedforward ANC system is to make $e_a(t)$ as close as possible to some desired signal $d_a(t)$. Specifically, the feedforward ANC *problem* can be stated as follows:

Find the single-input/single-output system H_a that minimizes

$$J\left(\{[x_a(\tau) + (GoH_a)(q_a(\tau)), d_a(\tau)] : \; 0 \le \tau \le T\}\right). \qquad (2.37)$$

Comparing the feedforward signal estimation problem to the feedforward ANC problem, we see that the feedforward ANC problem can be viewed as a constrained version of the feedforward signal estimation problem. Specifically, the feedforward ANC problem can be stated as finding H_s that minimizes the expression in eq. (2.36) subject to the constraint that H_s must be of the following form

$$H_s = GoH_a, \qquad (2.38)$$

for a given system G and some system H_a.

We now compare the single-channel feedforward signal estimation system to the single-channel feedforward ANC system in terms of their input/output characteristics. To this end, we assume that both systems are driven by the same inputs and that the desired signal for both systems is the same. We also assume that each system is designed so that the the distance between its output and the desired signal is minimized.

Under these assumptions, the performance of the ANC system is never better than the performance of the corresponding signal estimation system, since the optimal ANC system is obtained by minimizing $J(.)$ subject to a constraint, and the optimal signal estimation system is obtained by minimizing $J(.)$ without any constraints.

Again under these assumptions, if the minimizing H_s happens to equal (GoQ) for some system Q, the optimal ANC system will be identical to the optimal signal estimation system. Specifically in this case, the minimizing H_a' in the ANC problem is system Q. For example, if G is invertible, the minimizing H_s can be expressed as

$$H_s = IoH_s \tag{2.39}$$

$$= Go(G^{-1}oH_s). \tag{2.40}$$

This implies that the solution to the feedforward ANC problem can be expressed in terms of the solution to the corresponding feedforward signal estimation problem with $H_a = G^{-1}oH_s$.

The above discussion implies that unless the solution to the signal estimation problem is (GoH_a) with H_a' being the solution to the corresponding ANC problem, the optimal ANC system will be different from the optimal signal estimation system.

In general, a procedure for solving the feedforward signal estimation problem can not be used to solve the feedforward ANC problem, since a procedure for solving an un-constrained minimization problem cannot be used to solve a constrained minimization problem.

If the minimizing H_a in the ANC system can be commuted with G, the feedforward ANC problem can be solved by solving a related feedforward signal estimation problem. Specifically in this case, the minimizing H_a is the solution to the signal estimation problem with $x_s(t) = x_a(t)$ and $q_s(t) = G(q_a(t))$.

2.6 Summary

Acoustical principles behind ANC and several commonly used control strategies for ANC were reviewed in this chapter. The relationship between active noise cancellation and signal estimation was also explored.

Numerical studies were presented to demonstrate the feasibility of minimizing the total acoustic potential energy in an enclosure. These studies showed that at low frequencies, a few secondary sources can be used to obtain substantial reduction in the total acoustic potential energy in an enclosure provided that two or three modes dominate the response of the enclosure. In another study, it was found that driving the acoustic pressure to zero at a single point in an enclosure produces a zone of quiet around this point which has a diameter of about one tenth of the wavelength of the excitation frequency. It is important to recall that the secondary source strengths in all these studies were calculated based on the exact knowledge of the primary source strength for all time. However in a practical ANC system, the primary source strength is unknown and must be estimated based on the available measurements.

Existing control strategies for ANC were characterized as pointwise feedforward controllers or as pointwise feedback controllers.

Feedforward systems achieve noise cancellation by exploiting the cross-correlation between the reference signal $q(t)$ and the unwanted noise $n(t)$ at the error microphone. For the single-channel feedforward ANC system, we showed that the optimal, in the minimum mean-square sense, canceling signal at the error microphone is the LLSE of $-n(t)$ based on

$$p''(t) = q(t) * g_{all}(t), \qquad (2.41)$$

where $G_{all}(s)$ is the all-pass part of the transfer function from the input of the canceling loudspeaker to the value of the canceling field at the location of the error microphone, see equation (2.28). We also showed that this formulation can be used to compute the optimal performance of the feedforward system based on the second-order statistics of $\{n(t), q(t)\}$.

Feedback systems achieve noise cancellation by exploiting the auto-correlation of the noise $n(t)$ at the error microphone. For the single-channel feedback ANC system, we showed that the optimal canceling signal at the error microphone is the LLSE of $-n(t)$ based on

$$p'(t) = n(t) * g_{all}(t), \qquad (2.42)$$

where $G_{all}(s)$ is the all-pass part of the transfer function from the input of the canceling loudspeaker to the output of value of the canceling field at the location of the error microphone, see equation (2.26). We also showed that this formulation can be used to compute

51

the optimal performance of the feedback system based on the second-order statistics of $n(t)$.

Using the above analysis we can compare the optimal performance of the feedforward system to the optimal performance of the feedback system. Based on this analysis, we expect the feedforward system to achieve low levels of attenuation whenever the reference signal is weakly correlated with the noise at the error microphone.

The main drawback of the feedforward systems is the problem of how to acquire a reference signal that is highly correlated with the unwanted noise and at the same time avoid acoustic feedback from the output of the canceling loudspeaker to the reference signal.

The main drawback of analog feedback systems is that these systems can not be tuned to take advantage of the statistical characteristics of the unwanted noise. The main drawback of the digital feedback system of Graupe [26] is that it can only be used when the transfer function from the canceling loudspeaker to the error microphone contains no delay. This severely limits the applications in which this system can be used. Furthermore, a multiple-microphone version of the system proposed by Graupe does not exist.

In Section 2.5, we explored the relationship between signal estimation and active noise cancellation. The systems studied in this Section were free to be linear or non-linear and the cost functions involved were quite general. By structuring the signal estimation problem in a particular way, we were able to illustrate the close relationship between single-channel signal estimation and single-channel feedback active noise cancellation and the close relationship between feedforward signal estimation and and feedforward active noise cancellation. In each case, we showed that the active noise cancellation problem can be viewed as a constrained version of the signal estimation problem.

In the context of pointwise active noise cancellation, our goal is to develop a general framework for designing feedback active noise cancellation systems that can be used in a wide range of applications. Our formulation can be used in single-microphone as well as multiple-microphone configuration, and the resulting algorithms will take advantage of the the statistical characteristics of the unwanted noise.

We will then extend the above framework to the case where the performance criterion is distributed. The performance criterion in this case is the total acoustic energy in an

52

enclosure. However, the measurements are still made using ordinary microphones, and the control signal is still used to drive an ordinary loudspeaker. Most importantly, the control signals are generated causally based only on the outputs of the microphones.

Chapter 3

Pointwise Optimal Feedback Control for ANC

3.1 Introduction

In this chapter we develop an optimal feedback controller for attenuating the acoustic pressure at the locations of a finite number of error microphones. Our controller is designed to achieve the same acoustic objective as the multiple-channel feedforward controller in [18]; however, our controller generates the control signal based on the outputs of the error microphones, and not based on a separate reference signal. Therefore, the feedback ANC system developed in this chapter achieves the same acoustic objective as the feedforward ANC system in [18], without the problems associated with acquiring an appropriate reference signal.

An important feature of the optimal feedback controller is that it takes full advantage of the statistics of the noise. LMS-type feedforward controllers used in ANC systems take advantage of the statistical characteristics of the noise to be canceled. However, existing feedback ANC controllers do not take advantage of the noise statistics (except under very restrictive assumptions [26]).

We present a procedure for designing optimal feedback controllers for pointwise ANC that can be used under very general conditions. The error microphones are placed at those locations where noise attenuation is desired, and the canceling loudspeaker is placed in the vicinity of these microphones. Given a specific arrangement of the error microphones and the canceling loudspeaker, and given a specific statistical model for the noise at these microphones, our design procedure determines the optimal feedback controller. Specifically, our design procedure finds the controller that minimizes a performance criterion involving the mean-square sum of the residual signals at the error microphones and the mean-square sum of the control signal.

Our approach is to model the residual signals at the microphones as the sum of the outputs of two linear system. The unwanted noise signals at the microphones are modeled as the outputs of a multiple-input/multiple-output linear system driven by a white process. Similarly, the canceling signals at the microphones are modeled as the outputs of a single-input/multiple-output linear system driven by the control signal. Finally, the residual signals at the microphones are modeled as the sum of the outputs of these two linear systems. We show that the control signal that minimizes a certain class of pointwise performance criteria is a linear feedback of the estimated states of these two linear systems. These state estimates are computed using a Kalman filter, and the feedback gain is computed by iterating a Riccati equation. The single-microphone case is presented first, and the results are then extended to the multiple-microphone case.

In Section 3.4, we show that the optimal feedback controller developed in this chapter can be thought of as a generalization of the optimal feedforward controller. Specifically, the optimal feedforward controller generates the canceling signal based on all the available measurement, but the optimal feedforward controller generates the canceling signal based on a proper subset of the available measurements.

The control algorithms developed in this chapter are discrete-time algorithms. Unless stated otherwise, we assume throughout this chapter that the noise to be attenuated, the canceling signal, and all other intermediate signals are appropriately sampled. Unless stated otherwise, throughout this chapter "t" represents the normalized sampling time.

3.2 Single-Microphone Optimal Feedback Controller for ANC

A single-microphone optimal feedback controller for ANC is developed in this Section. The system utilizes one error microphone and one canceling loudspeaker. The microphone is placed at the point where noise attenuation is desired, and the loudspeaker is placed as close as possible to this microphone. The acoustic objective is to attenuate the pressure at the location of the microphone. The loudspeaker produces a secondary field that interferes destructively with the primary field at the microphone, generating a zone of quiet around the microphone as reported in [15].

A generic single-microphone feedback ANC system is depicted in Fig. 3.1 where $m(t)$ is the microphone output, and $r(t)$ is the input of the canceling loudspeaker. The microphone measures the sum of the unwanted noise $n(t)$ and the canceling signal $c(t)$. The objective is to generate $r(t)$, based on the measurements of $m(t)$, so that the power of the residual signal at the microphone is minimized.

Throughout this chapter we rely on the following model for the microphone. The output of the microphone is assumed to be the sum of the acoustic pressure at the location of the microphone and the microphone measurement noise. According to this model, the only distortion introduced by the microphone is this additive measurement noise. This microphone model is quite accurate for most practical pressure microphones at frequencies below 1000Hz [6].

The block diagram for the single-microphone ANC system that we propose and develop in this Section is depicted in Fig. 3.2. The system $G(z)$ represents the overall transfer function from the canceling loudspeaker input $r(t)$ to the value of the canceling field at the location of the microphone $c(t)$ and incorporates the transfer function for the loudspeaker and the propagation path between the loudspeaker and the microphone. The microphone output $m(t)$ is the sum of the unwanted noise $n(t)$, the canceling signal $c(t)$, and the microphone measurement noise $v(t)$.

Our basic strategy for solving the ANC problem is to view this problem as a stochastic optimal control problem with $n(t)$ being the stochastic disturbance and $G(z)$ being the plant. From this point of view, the objective is to keep the output of the plant, $e(t) = c(t) + n(t)$, as close to zero as possible. Note that $e(t)$ is the value of the residual acoustic field at the location of the microphone, i.e. the sum of the noise field at the location of the microphone and canceling field at the location of the microphone.

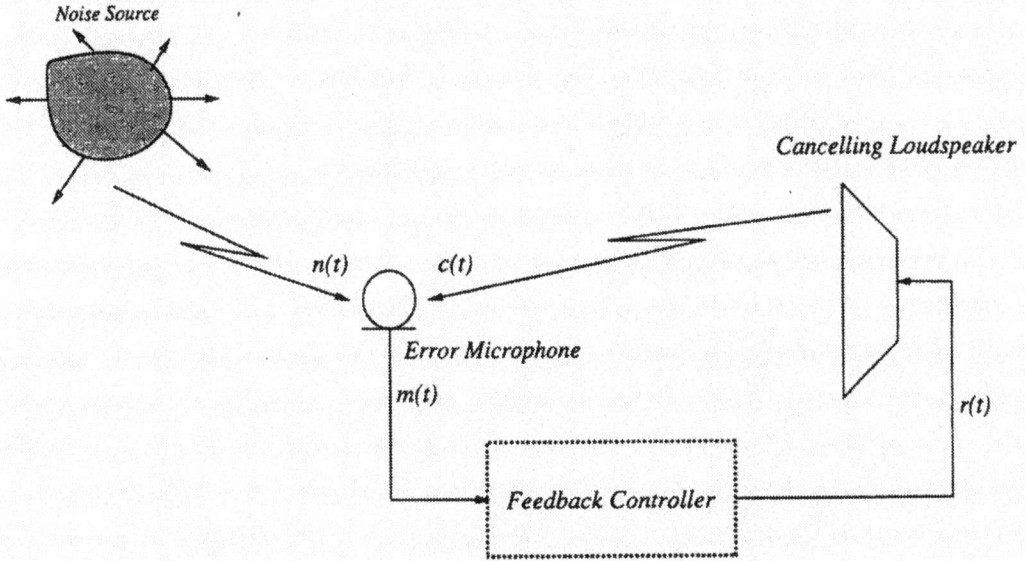

Figure 3.1: Generic single-microphone feedback active noise cancellation system.

3.2.1 Model Specification

The stochastic optimal control formulation requires a disturbance model and a plant model. We model the noise $n(t)$ as the output of a pole-zero system function driven by a white process:

$$n(t) = -\sum_{k=1}^{q} \alpha_k n(t-k) + \sum_{k=1}^{j} \beta_k w(t-k) \qquad j \leq q, \qquad (3.1)$$

where $w(t)$ is a zero mean unit variance white Gaussian process, i.e. $n(t)$ is modeled as an ARMA process. The plant model, corresponding to the transfer function from the input of the canceling loudspeaker to the value of the canceling field at location of the error microphone, is the pole-zero system function

$$G(z) = \frac{\sum_{k=1}^{m} b_k z^{-k}}{1 + \sum_{k=1}^{n} a_k z^{-k}} \qquad m \leq n. \qquad (3.2)$$

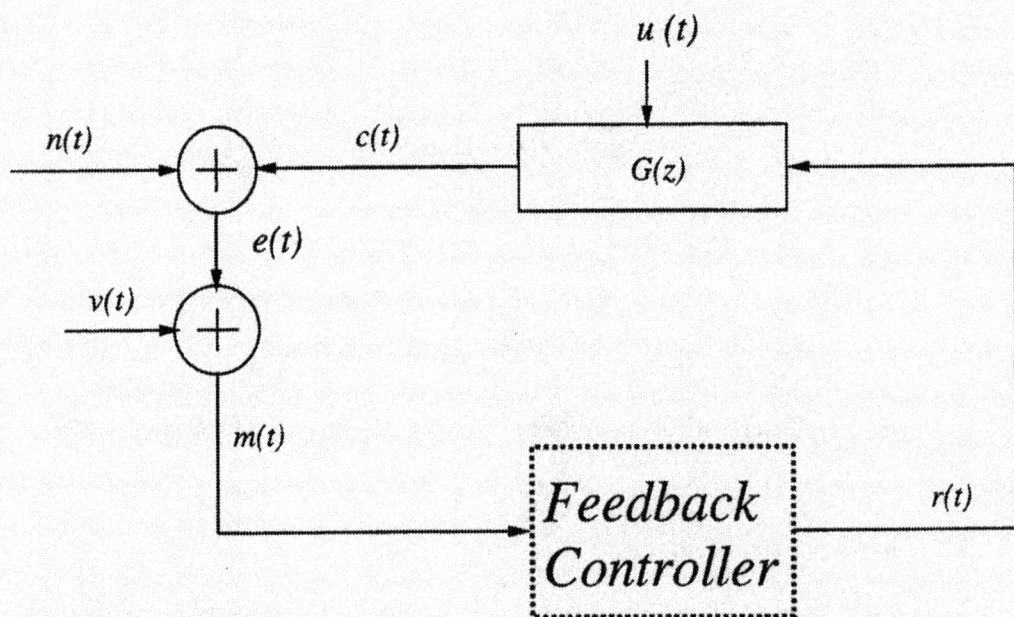

Figure 3.2: Block diagram of the single-microphone feedback active noise cancellation system.

We represent the system of Fig. 3.2 in state-space form in three steps. First we will derive a state-space representation for the disturbance model eq. (3.1), and then we will derive a state-space representation for the plant model eq. (3.2). These two representations are finally combined to get the overall state-space representation for the system in Fig. 3.2.

The disturbance model can be expressed in state-space form as [34, 3, 2].

$$x_d(t) = \Phi_d \, x_d(t-1) + L_d \, w(t-1) \tag{3.3}$$

$$n(t) = H_d \, x_d(t), \tag{3.4}$$

where $x_d(t)$ is the $(q \times 1)$ state vector for the disturbance model; Φ_d is the corresponding $(q \times q)$ transition matrix

$$\Phi_d = \begin{bmatrix} -\alpha_1 & 1 & 0 & \cdots & 0 \\ -\alpha_2 & 0 & \ddots & & \\ \vdots & \vdots & & \ddots & \\ -\alpha_{q-1} & 0 & \cdots & \cdots & 1 \\ -\alpha_q & 0 & \cdots & \cdots & 0 \end{bmatrix}; \tag{3.5}$$

L_d is the $(q \times 1)$ vector

$$L_d = [\beta_1 \; \cdots \; \beta_j \; 0 \cdots 0]^T;$$

and H_d is the $(1 \times q)$ unit vector

$$H_d = [1 \; 0 \cdots 0].$$

Similarly, the plant model can be expressed in state-space form as [34]:

$$x_p(t) = \Phi_p \, x_p(t-1) + G_p \, r(t-1) + L_p \, u(t-1) \tag{3.6}$$

$$c(t) = H_p \, x_p(t), \tag{3.7}$$

where $x_p(t)$ is the $(n \times 1)$ state vector for the plant model; $u(t)$ is an $(n \times 1)$ vector of random variables such that

$$E\left\{u(t)u(\tau)^T\right\} = \delta(t - \tau)I_{n \times n},$$

59

with $u(t)$ independent of $w(t)$; Φ_p is the $(n \times n)$ transition matrix of the plant

$$\Phi_p = \begin{bmatrix} -a_1 & 1 & 0 & \cdots & 0 \\ -a_2 & 0 & \ddots & & \\ \vdots & \vdots & & \ddots & \\ -a_{n-1} & 0 & \cdots & \cdots & 1 \\ -a_n & 0 & \cdots & \cdots & 0 \end{bmatrix}; \tag{3.8}$$

G_p is the $(n \times 1)$ vector

$$G_p = [b_1 \cdots b_m \ 0 \cdots 0]^T;$$

L_p is the $(n \times n)$ matrix

$$L_p = \sigma_u I_{n \times n};$$

and H_p is the $(n \times 1)$ unit vector

$$H_p = [1 \ 0 \cdots 0]^T.$$

The state-space model for the overall system is now obtained by combining (3.3-3.4) with (3.6-3.7):

$$x(t) = \Phi \, x(t-1) + G \, r(t-1) + L \begin{bmatrix} w(t-1) \\ u(t-1) \end{bmatrix} \tag{3.9}$$

$$m(t) = H x(t) + v(t), \tag{3.10}$$

where $x(t)$ is the state vector of the overall system

$$x(t) = \begin{bmatrix} x_d(t) \\ x_p(t) \end{bmatrix}; \tag{3.11}$$

Φ is the corresponding $(q+n) \times (q+n)$ transition matrix

$$\Phi = \begin{bmatrix} \Phi_d & 0 \\ 0 & \Phi_p \end{bmatrix}; \tag{3.12}$$

60

G is the $(q + n) \times 1$ vector

$$G = \begin{bmatrix} 0 \\ \vdots \\ 0 \\ G_p \end{bmatrix};$$

(3.13)

L is the $(q + n) \times (q + n)$ matrix

$$L = \begin{bmatrix} L_d & 0 \\ 0 & L_p \end{bmatrix};$$

and H is the $1 \times (q + n)$ vector

$$H = [H_d \ H_p].$$

The measurement noise in the error microphone is modeled by the white Gaussian process $v(t)$ whose variance is σ_v^2, and $v(t)$ is independent of $w(t)$ and $u(t)$.

3.2.2 The Optimal Feedback Controller

The goal of the ANC system is to choose $\{r(t) : t = 0, \ldots, N - 1\}$ so that the following performance measure is minimized:

$$J(N) = E\left\{ \sum_{t=0}^{N-1} \{e^2(t) + \rho \ r^2(t)\} + e^2(N) \right\}.$$

(3.14)

To be causally realizable, the choice for $r(t)$ must be a function of the available measurements at time "t", i.e. it must be a function of $\{m(\tau) : \tau \leq t - 1\}$. The first summand in (3.14) represents the power of the residual signal at the microphone, and the second summand represents the power expended by the control signal. Here, ρ is a weighting factor which is chosen to achieve a trade off between control effort and noise attenuation. Recalling that

$$e(t) = H \ x(t),$$

(3.15)

we can rewrite $J(N)$ in terms of the state vector $x(t)$ as

$$J(N) = E\left\{ \sum_{t=0}^{N-1} \{x^T(t)H^T H x(t) + \rho \ r^2(t)\} + x^T(N)H^T H x(N) \right\}.$$

(3.16)

61

It is important to observe that (3.16) is the expected value of a quadratic function of the state vector $x(t)$ and the control signal $r(t)$. Following standard terminology from control literature, we shall refer to this minimization problem as the "Finite Horizon Optimal Linear Quadratic Control Problem".

Since the performance measure $J(N)$ is the expected value of a quadratic form in the state and the control, the minimizing choice for $\{r(t) : t = 0, \ldots, N-1\}$ is a linear feedback of the optimal state estimates [24]. The optimal state estimates can be computed using a Kalman filter, and the feedback gain can be computed by iterating a Riccati difference equation. To this end, let us denote by

$$\hat{x}(t|t) = E\{x(t)|m(1), \ldots, m(t)\} \tag{3.17}$$

the state estimate based on observations up to time "t", and by

$$P(t|t) = E\left\{[\hat{x}(t|t) - x(t)][\hat{x}(t|t) - x(t)]^T | m(1), \ldots, m(t)\right\} \tag{3.18}$$

the associated error covariance matrix.

Then using the standard Kalman filter formulation, $\hat{x}(t|t)$ and $P(t|t)$ can be computed sequentially in time, in two stages, as follows:

Propagation Equations:

$$\hat{x}(t|t-1) = \Phi\,\hat{x}(t-1|t-1) + G\,u(t-1) \tag{3.19}$$

$$P(t|t-1) = \Phi\,P(t-1|t-1)\,\Phi^T + L\,L^T \tag{3.20}$$

Updating Equations:

$$\hat{x}(t|t) = \hat{x}(t|t-1) + K_f(t)[m(t) - H\,\hat{x}(t|t-1)] \tag{3.21}$$

$$P(t|t) = [I - K_f(t)\,H]P(t|t-1), \tag{3.22}$$

where $K_f(t)$ is the Kalman gain given by

$$K_f(t) = \frac{1}{H\,P(t|t-1)H^T + \sigma_v^2}P(t|t-1)H^T. \tag{3.23}$$

The minimizing choice for $\{r(t): \ t = 0, \ldots, N-1\}$ is then given by

$$r(N-j) = -[GP(j-1)G + \rho]^{-1}G^T P(j-1)\Phi\,\hat{x}(N-j|N-j) \qquad j = 1, \ldots, N$$

$$= K_c(j-1)\,\hat{x}(N-j|N-j), \tag{3.24}$$

where the $(m + n) \times (m + n)$ matrix $P(j)$ is obtained by iterating the following Riccati difference equation:

$$P(j + 1) = \Phi^T P(j)\Phi - \Phi^T P(j)G[G^T P(j)G + \rho]^{-1}G^T P(j)\Phi + H^T H, \qquad (3.25)$$

with the initial condition $P(0) = H^T H$. The resulting minimum value of the performance index $J(N)$ can also be computed based on $P(N)$ [35]. The block diagram for the resulting ANC system is depicted in Figure 3.3.

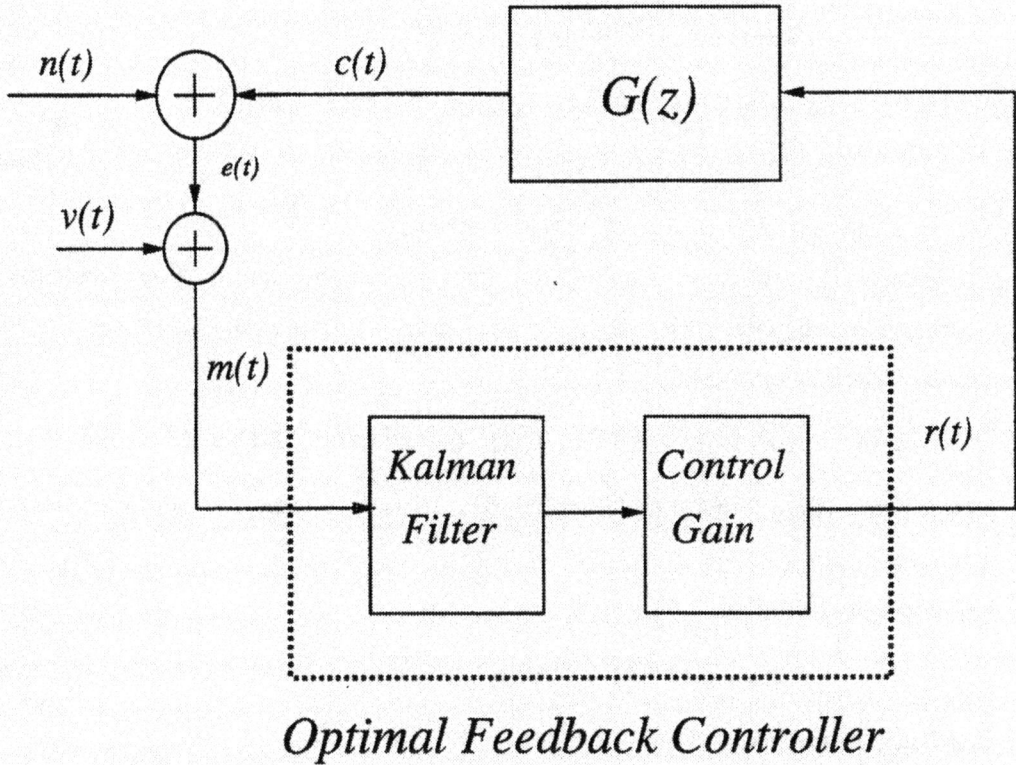

Figure 3.3: Block diagram of our single-microphone feedback ANC system.

Infinite Horizon Case

If we consider the above problem with both time-invariant system matrices and with constant weighting factor ρ, and if we allow N to approach infinity, a new performance index

63

can be defined as

$$J = \lim_{N \to \infty} \frac{1}{N} J(N).$$

In this case, we ask for the choice of $\{r(t) : t = 0, 1, \ldots\}$ that minimizes J. This is a well posed problem provided that $[\Phi, G]$ is stabilizable, and $[\Phi, H]$ is detectable. In this case, $P(j)$ converges to a constant matrix $P(\infty)$ which is the unique positive definite solution of the following algebraic Riccati equation

$$P(\infty) = \Phi^T P(\infty) \Phi - \Phi^T P(\infty) G [G^T P(\infty) G + \rho]^{-1} G^T P(\infty) \Phi + H^T H. \qquad (3.26)$$

In this case, the minimizing choice for $r(t)$ is given by

$$r(t) \;=\; -[G^T P(\infty) G + \rho]^{-1} G^T P(\infty) \Phi \; \hat{x}(t|t) \qquad (3.27)$$

$$\;=\; K_c(\infty) \; \hat{x}(t|t). \qquad (3.28)$$

Note that the control gain, $K_c(\infty)$, is time-invariant.

Similarly, the error covariance matrix, $P(t|t-1)$ in the Kalman filter, converges to a constant matrix $P(\infty)$ which is the unique positive definite solution of the following algebraic Riccati equation

$$P(\infty) = \Phi P(\infty) \Phi^T - \Phi P(\infty) H^T [H P(\infty) H^T + \sigma_v^2]^{-1} H P(\infty) \Phi^T + L L^T,$$

provided that $[\Phi, L]$ is stabilizable and $[\Phi, H]$ is detectable. The Kalman filter equations reduce to:

Propagation Equation:

$$\hat{x}(t|t-1) = \Phi \; \hat{x}(t-1|t-1) + G \, u(t-1) \qquad (3.29)$$

Updating Equation:

$$\hat{x}(t|t) = \hat{x}(t|t-1) + K_f(\infty)[m(t) - H \; \hat{x}(t|t-1)], \qquad (3.30)$$

where $K_f(\infty)$ is the Kalman gain given by

$$K_f(\infty) = \frac{1}{H P(\infty) H^T + \sigma_v^2} P(\infty) H^T. \qquad (3.31)$$

Note that the Kalman filter in this case is linear and time-invariant. Recalling that the control gain $K_c(\infty)$ is also time-invariant, we conclude that the controller becomes a linear time-invariant system in this case.

64

3.2.3 Frequency-Domain Perspective

Our formulation of the ANC problem so far has been entirely in the time domain. Additional insight can be gained by looking at the closed-loop ANC system in the frequency domain. For this purpose, we shall concentrate on the infinite horizon case in which the overall system is linear and time-invariant, and all the signals involved are jointly stationary. Furthermore, we assume that the objective is to minimize the pressure fluctuations at the error microphone, i.e. we assume that the weighting factor for control effort, ρ in eq. (3.14), is zero.

Let $H(z)$ denote the transfer function of the controller in Fig. 3.3. Since the measurement noise $v(t)$ is white and $G(z)$ contains at least one delay, the causal controller that minimizes $E\{m^2(t)\}$ is the same as the causal controller that minimizes $E\{e^2(t)\}$ (note that $m(t) = e(t) + v(t)$). Therefore, we can characterize the optimal feedback controller as the controller that minimizes $E\{m^2(t)\}$ in Fig. 3.3. Referring to this figure, we define the closed-loop transfer function $T(z)$ as

$$T(z) = \frac{1}{1 - H(z)\,G(z)} \tag{3.32}$$

, and we define $z(t)$ as

$$z(t) = n(t) + v(t). \tag{3.33}$$

The power spectrum of the microphone output $P_{mm}(e^{j\omega})$ can be expressed in terms of the power spectrum of $z(t)$, i.e.

$$P_{mm}(e^{j\omega}) = \left|T(e^{j\omega})\right|^2 P_{zz}(e^{j\omega}). \tag{3.34}$$

Equations (3.32)-(3.34) show how the controller $H(e^{j\omega})$ affects the power spectrum of the microphone output. To minimize the variance of $m(t)$ or equivalently to minimize the variance of $e(t)$, we need to make $|T(e^{j\omega})|$ small over those frequencies for which the unwanted noise has significant power. This is precisely what the optimal controller does. Specifically, the optimal controller uses the noise model to determine those frequencies for which the unwanted noise signal has significant power, and it chooses $H(z)$ so that the magnitude of the resulting closed-loop transfer function, $|T(e^{j\omega})|$, is small over these

65

frequencies. From this point of view, it is clear that the more narrow-band the spectrum of $n(t)$, the higher the attenuation level achieved by the feedback controller. An example of this will be presented in the next section.

Having a good model for the unwanted noise is crucial for achieving high attenuation levels. In the absence of an accurate model for the the unwanted noise (i.e. without knowing at which frequencies the power of the unwanted noise is concentrated), one is forced to assume that the spectrum of the noise is flat over a wide range of frequencies. This results in an ANC system with low attenuation over a wide range of frequencies (e.g., noise canceling headphones developed by Wheeler [51]). On the other hand, in cases for which the unwanted noise is relatively narrow band and the noise is well modeled, the active noise cancellation system can be designed to achieve very high attenuation levels. This is done by choosing the controller such that the closed-loop system has high attenuation over the narrow band of frequencies where the power of the noise is concentrated.

Analysis of the feedback controller in chapter 2 can also be used to explain why relatively narrow band noise is highly attenuated by the optimal feedback controller. Recall that this analysis was done in continuous time; hence, "t" represent the continuous time variable in this paragraph. Referring to Fig. 2.3 of Chapter 2, it was shown that the optimal canceling signal in the single-channel feedback ANC system is the LLSE of $-n(t)$ based on

$$p'(t) = n(t) * g_{all}(t), \tag{3.35}$$

where $G_{all}(s)$ is the all-pass part of $G(s)$. The mean-square estimation error will be small whenever $n(t)$ can be estimated accurately based on $p'(t)$. For a relatively narrow band $n(t)$, $G_{all}(s)$ can be approximated as a delay. Therefore in the narrow band case, the problem reduces to predicting $n(t)$ based on $p'(t) = n(t - \tau)$, where τ is the delay. More importantly, the mean-square estimation error will be small with a narrow band $n(t)$, since a narrow band signal is highly predictable.

3.2.4 Frequency-Weighted Optimal Controller for ANC

Assuming time-invariant system matrices and stationary acoustic noise, we shall drive a feedback controller that minimizes a cost function involving the frequency-weighted power

66

of the residual signal at the microphone. Referring to Fig. 3.2, the objective in this section is to minimize

$$J_\infty = \lim_{N \to \infty} \frac{1}{N} E \left\{ \sum_{t=0}^{N-1} \left(e_w^2(t) + \rho\, r^2(t) \right) + e_w^2(N) \right\},$$ (3.36)

where $e_w(t) = e(t) * s(t)$, with $s(t)$ being a pre-specified weighting filter. The relationship between $e(t)$ and $e_w(t)$ can be expressed in state-space form as:

$$x_w(t) = \Phi_w\, x_w(t-1) + G_w e(t)$$ (3.37)

$$e_w(t) = H_w\, x_w(t).$$ (3.38)

The weighting filter $s(t)$ might represent the perceived loudness of sound at different frequencies. In this context, the filter $s(t)$ is typically referred to as an A-weight filter.

The state-space representation eqs. (3.37)-(3.38) of the weighting filter can be combined with the state-space representation eqs. (3.9)-(3.10) of the original acoustic system. The state-space representation for the combined system is

$$x_c(t) = \Phi_c\, x_c(t-1) + G_c\, r(t-1) + L_c \begin{bmatrix} w(t-1) \\ u(t-1) \end{bmatrix}$$ (3.39)

$$m(t) = H_c\, x_c(t) + v(t),$$ (3.40)

where $x_c(t)$ is the state vector of the combined system

$$x_c(t) = \begin{bmatrix} x(t) \\ x_w(t) \end{bmatrix};$$ (3.41)

Φ_c is the corresponding transition matrix

$$\Phi_c = \begin{bmatrix} \Phi & 0 \\ G_w H & \Phi_w \end{bmatrix};$$ (3.42)

G_c is the column vector

$$G_c = \begin{bmatrix} G \\ o \end{bmatrix};$$ (3.43)

L_c is the matrix

$$L_c = \begin{bmatrix} L \\ \mathbf{o} \end{bmatrix} ; \qquad (3.44)$$

and H_c is the row vector

$$H_c = [H \quad \mathbf{o}]. \qquad (3.45)$$

Next, we express the performance criterion J_∞ in terms of the state vector $x_c(t)$. For this purpose, we define

$$J(N) = E\left\{ \sum_{t=0}^{N-1} \left(e_w^2(t) + \rho\, r^2(t) \right) + e_w^2(N) \right\}. \qquad (3.46)$$

Recalling that

$$e_w(t) = [\mathbf{o} \quad H_w]\, x_c(t), \qquad (3.47)$$

we can rewrite $J(N)$ in terms of the state vector $x_c(t)$ as

$$J(N) = E\left\{ \sum_{t=0}^{N-1} \left(x_c^T(t) C_\star^T C_\star x_c(t) + \rho\, r^2(t) \right) + x_c^T(N) C_\star^T C_\star x_c(N) \right\}, \qquad (3.48)$$

where $C_\star = [\mathbf{o} \quad H_w]$. Note that $J(N)$ is the expected value of a quadratic function of the state vector $x_c(t)$ and the control signal $r(t)$.

Since $J_\infty = \lim_{N \to \infty} \frac{1}{N} J(N)$, the choice of $\{r(t) : t \geq 0\}$ that minimizes J_∞ is a linear feedback of the optimal state estimates of the combined system, provided that $[\Phi_c, G_c]$ is stabilizable and $[\Phi_c, H_c]$ is detectable. The exact equations for the controller are obtained following steps identical to the ones followed to get the infinite horizon controller in Section 3.2.2.

3.2.5 Implementation and Performance of the Optimal Feedback Controller

Two sets of experiments were performed to evaluate the performance of the optimal single-microphone feedback ANC system. In the first set, computer simulations on recorded aircraft noise were used to compare the performance of the optimal feedback system to the performance of the feedforward system proposed in [10]. The second set of experiments were carried out to measure the performance of the hardware implementation of the optimal feedback system in the context of aircraft noise. The performance of the hardware implementation of the optimal feedback system was also compared to the performance of a commercial analog feedback system built by the BOSE Corporation.

Our feedback system and the feedforward system in [10] were evaluated using a set of noise canceling headphones. These headphones were equipped with a reference microphone, an error microphone, and a canceling loudspeaker. The reference signal for the feedforward controller was obtained as the output of the reference microphone that was attached to the outside of the headphones, and the error microphone was placed inside the headphones. The canceling loudspeaker was also placed inside the headphones, about 1.2 centimeters away from the error microphone. In this case, the transfer function $G(z)$ of Figure 3.2 corresponds to the transfer function from the canceling loudspeaker to the error microphone inside the headphones.

A separate set of experiments was performed initially to estimate the transfer function $G(z)$ and the coefficients of the disturbance model. It was found that an ARMA model with three zeros and four poles produces a very good approximation to the transfer function from the canceling loudspeaker to the error microphone. The disturbance model was an all-pole system function of order four, and the coefficients of this model ($\alpha'_k s$ in (3.1)) were estimated using the algorithm in [40]. All the controllers discussed in this section are designed based on this estimated plant model and this estimated disturbance model.

Simulations

The performance of the feedforward system of Burgess [10] was compared to the performance of the infinite horizon feedback controller developed in Section 3.2.2. In computer simulations, the feedforward system of Burgess was able to attenuate the noise generated by a propeller aircraft by about 17dB. This result was obtained with a 350-tap finite impulse response controller and using recordings of the noise made at the reference microphone and at the error microphone. The sampling rate was 6KHz. Using the same recorded noise at the error microphone, our infinite horizon feedback controller was able to attenuate the

69

noise power at the error microphone by about 26dB. Referring to Fig. 3.2, this attenuation was calculated according to the following equation:

$$\text{Un-Weighted Attenuation} = 10 \log_{10} \frac{E\{e^2(t)\}}{E\{n^2(t)\}}. \tag{3.49}$$

The feedforward controller and the feedback controller were designed based on the exact knowledge of the plant model and the disturbance model. The power spectrum of the unwanted noise and the power spectrum of the residual signal obtained with the infinite horizon feedback controller are depicted in Fig. 3.4(a).

Another set of simulations was performed to illustrate the performance of the frequency-weighted optimal feedback controller. Again, we used the same propeller aircraft noise that was used in the simulations described in the previous paragraph. The frequency response of the weighting filter, $s(t)$, that was used in the design of the controller is depicted in Fig. 3.5. This filter models the variations in perceived loudness of sound at different frequencies. According to this model, a pure tone of unit power at 1KHz sounds 70dB louder than a pure tone of unit power at 10Hz. The power spectrum of the unwanted noise at the microphone and the power spectrum of the residual signal obtained with the frequency-weighted optimal feedback controller are depicted in Fig. 3.4(b). The un-weighted attenuation in this case is about 20dB. Referring to Fig. 3.2, let us define the weighted attenuation as

$$\text{Weighted Attenuation} = 10 \log_{10} \frac{E\{e_w^2(t)\}}{E\{n_w^2(t)\}}, \tag{3.50}$$

where $e_w(t) = e(t) * s(t)$ and $n_w(t) = n(t) * s(t)$. The weighted attenuation in this case is about 16dB. For comparison, the weighted attenuation corresponding to Fig. 3.4(a) is about 11dB. We see that incorporation of the weighting filter in the design of the feedback controller improves the weighted attenuation by about 5dB.

Hardware Implementation

The infinite horizon feedback controller, designed without frequency weighting, was implemented in hardware using a single AT&T DSP32C chip with 6KHz sampling rate. The resulting noise canceling headphones were placed in an enclosure, and propeller aircraft noise was injected into this enclosure. The headphones were able to attenuate the noise power at the error microphone by 25dB.

The spectrum of the unwanted noise at the output of the error microphone and the spectrum of the attenuated noise at the output of the error microphone obtained with our feedback controller are depicted in Fig. 3.6(a). The closed-loop transfer function realized by our controller, $T(e^{j\omega})$, is depicted in Fig. 3.6(b). As expected from the discussion in Section 3.2.3, the closed-loop system has high attenuation over the 90Hz-130Hz band where most of the power of the unwanted noise signal is concentrated.

In Figure 3.7, the performance of our optimal feedback controller is compared to that of the analog feedback controller developed commercially by BOSE Corporation. The spectrum of the original noise at output of the error microphone is depicted in Fig. 3.7(a). The spectrum of the attenuated noise at the output of the error microphone obtained with our controller and the spectrum of the attenuated noise at the output of the error microphone obtained with the analog controller are depicted in Fig. 3.7(b). The overall attenuation obtained with the analog feedback system is 17dB, and the overall attenuation obtained with the optimal feedback system is 25dB. Looking at Fig. 3.7, we see that the optimal feedback controller achieves higher attenuation than the analog controller over the 90Hz-130Hz band where most of the energy of the unwanted noise is concentrated. On the other hand, the optimal feedback controller achieves lower attenuation than the analog controller over the 350Hz-700Hz band where the original noise does not have significant power.

The main drawback of the above formulation is that the disturbance model and the plant model must be known before the optimal controller can be computed.

3.3 Multiple-Microphone Optimal Feedback Controller for ANC

In this section, the optimal control framework of Section 3.2 is extended to develop a feedback controller for attenuating the noise at the locations of "l" error microphones. The error microphones are placed where noise cancellation is desired, and the canceling loudspeaker

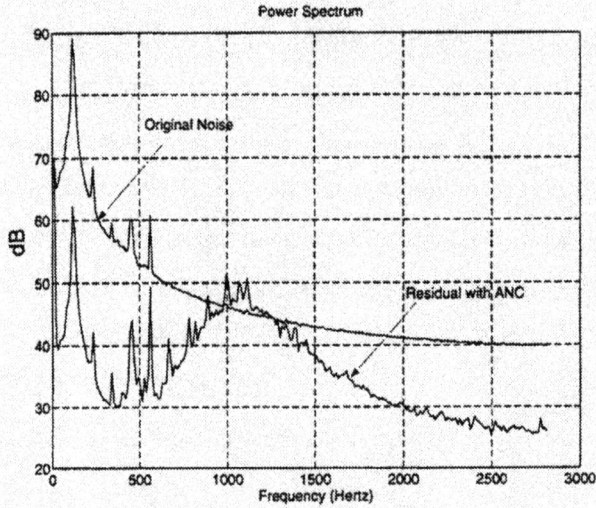

(a) Power spectrum of the propeller aircraft noise at the error microphone obtained using a feedback controller designed without a weighting filter.

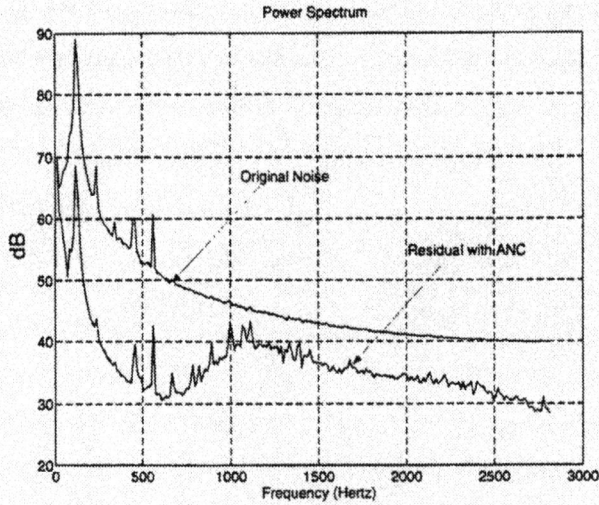

(b) Power spectrum of the proper aircraft noise at the error microphone obtained with a frequency-weighted feedback controller.

Figure 3.4: Results of our computer simulations.

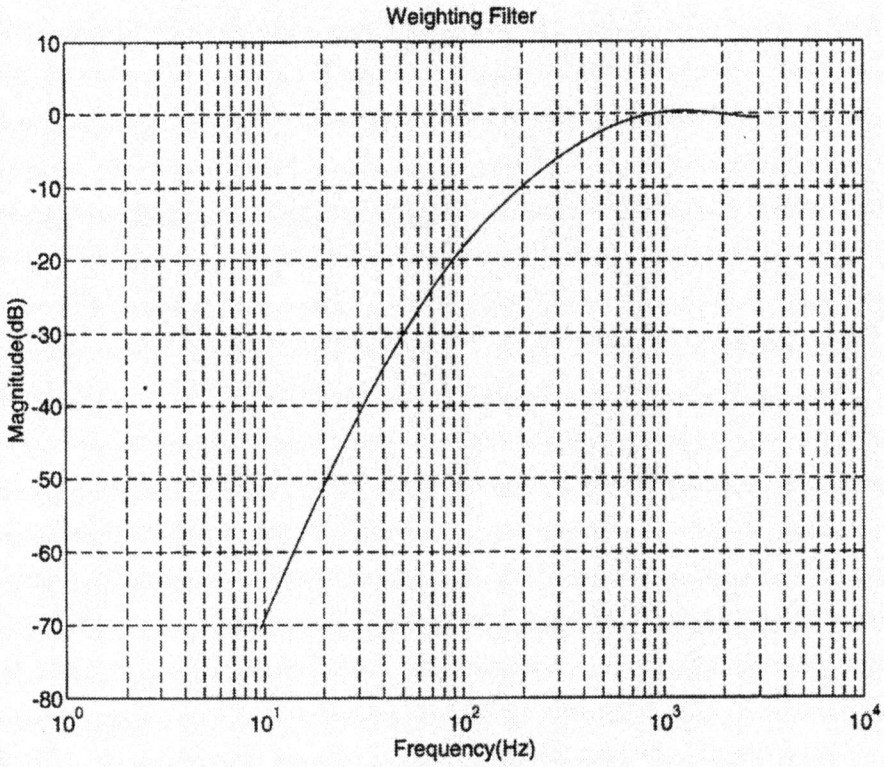

Figure 3.5: Frequency response of the weighting filter.

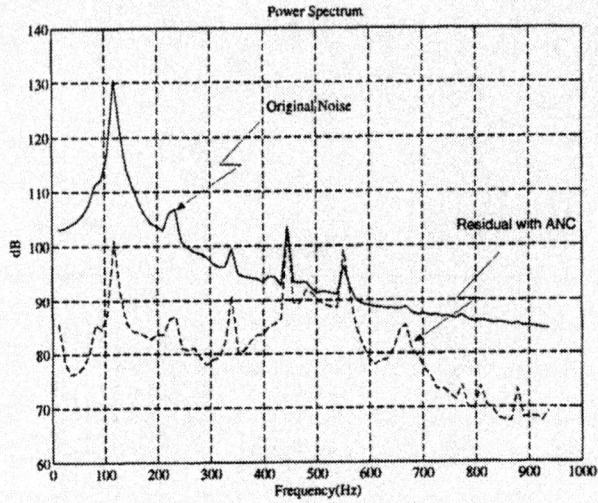

(a) Power spectrum of the output of the error microphone with and without the optimal feedback controller.

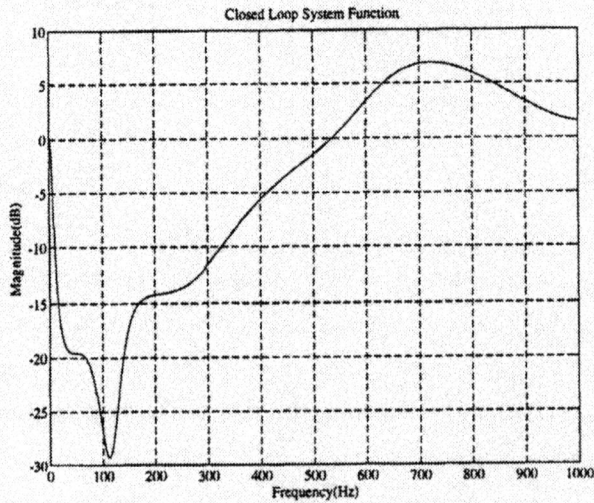

(b) Closed-loop transfer function $T(e^{j\omega})$ that is obtained by using the optimal feedback controller.

Figure 3.6: Performance of the hardware implementation of the optimal feedback controller.

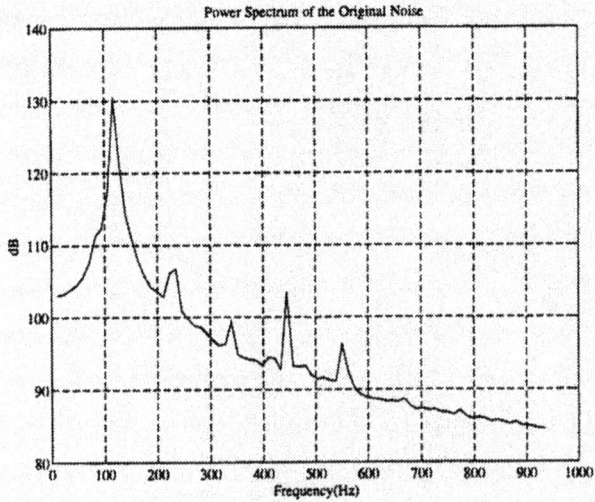

Power Spectrum of the Original Noise

(a) Power spectrum of the propeller aircraft noise at the error microphone.

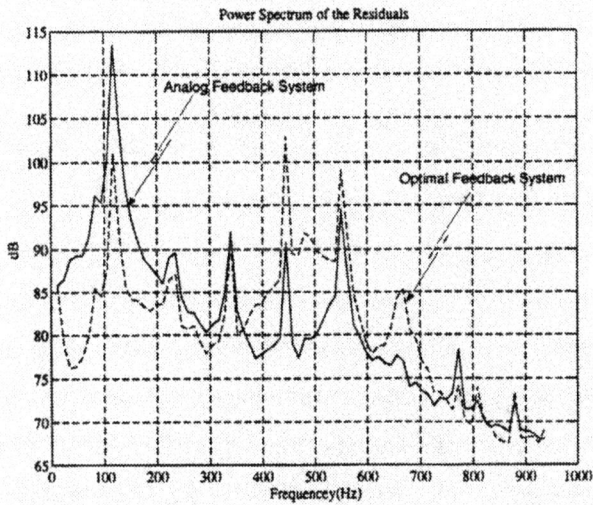

Power Spectrum of the Residuals

(b) Comparison of the analog feedback controller to the optimal feedback controller.

Figure 3.7: Performance of the hardware implementation of the optimal feedback controller.

is placed in the vicinity of these microphones. The acoustic objective is to attenuate the mean-square sum of the pressure fluctuations at the error microphones. The noise cancellation environment is depicted in Fi.g 3.8. Let us point out that our formulation can accommodate multiple canceling loudspeaker; however, for simplicity we will only consider the single-loudspeaker case in this section.

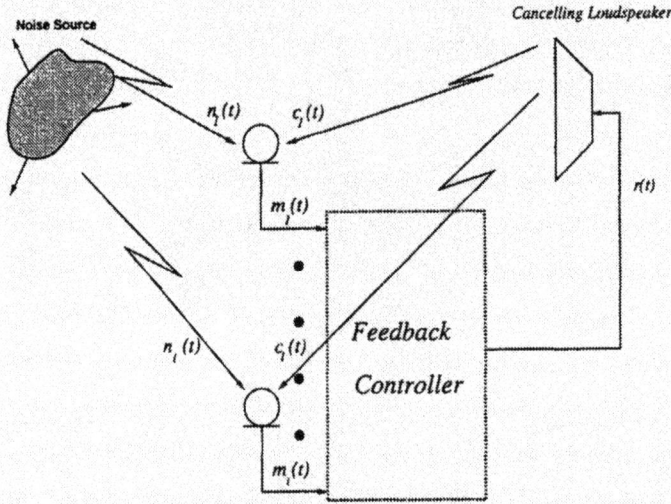

Figure 3.8: Multiple-microphone active noise cancellation system.

The block diagram corresponding to the system in Fig. 3.8 is depicted in Fig. 3.9. The plant model is a single-input/multiple-output linear system with input $r(t)$ and outputs $\{c_i(t)\}_{i=1}^{l}$. The output of the i-th microphone $m_i(t)$ is the sum of the unwanted noise at this microphone $n_i(t)$, the canceling signal at this microphone $c_i(t)$, and the measurement noise at this microphone $v_i(t)$. The unwanted noise signals $\{n_i(t)\}_{i=1}^{l}$ are modeled as the outputs of a multiple-input/multiple-output linear system driven by the white Gaussian vector-valued process $w(t)$, where $w(t) = \{w_i(t)\}_{i=1}^{l}$. The objective is to choose $\{r(t) : t = 0, \ldots, N-1\}$, based on the observations of $\{m_i(t)\}_{i=1}^{l}$, such that

$$J(N) = E\left\{ \sum_{t=0}^{N-1} \left(\left(\sum_{i=1}^{l} e_i^2(t)a_i \right) + \rho\, r^2(t) \right) + \sum_{i=1}^{l} e_i^2(N)a_i \right\} \qquad (3.51)$$

is minimized, where a_i is a positive weighting factor for noise cancellation at the i-th microphone, and ρ is a weighting factor for the overall control effort.

76

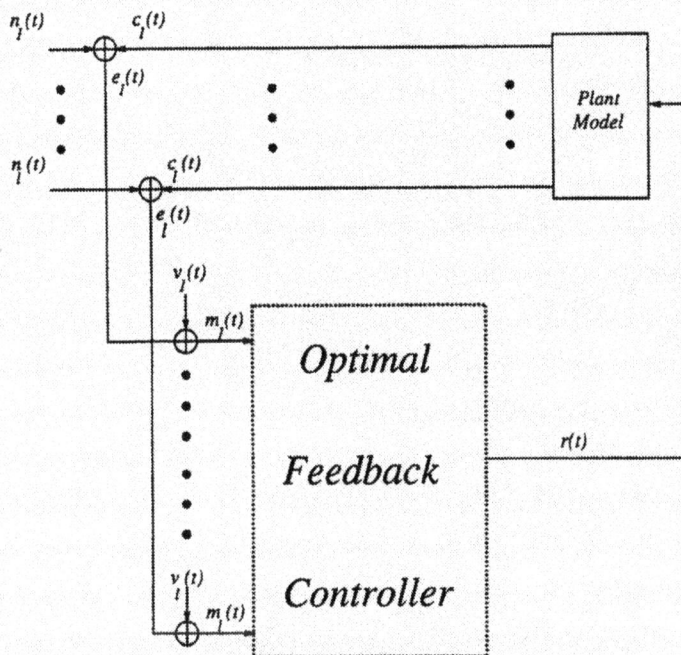

Figure 3.9: Block diagram of the multiple-microphone active noise cancellation system.

The unwanted noise signals at the microphones are modeled as the l outputs of the following linear system:

$$x_d(t) = \Phi_d x(t-1) + L_d w(t-1) \tag{3.52}$$

$$n(t) = H_d x_d(t-1), \tag{3.53}$$

where $n(t)$ is the ($l \times 1$) vector of the unwanted noise signals

$$n(t) = [n_1(t),\ n_2(t),\ \dots,\ n_l(t)]^T;$$

$x_d(t)$ is the state vector of this system, and the input of this system is the ($l \times 1$) vector $w(t)$ with

$$E\{w(t)w(\tau)^T\} = \delta(t-\tau)\,I.$$

The plant model is the following single-input/multiple-output linear system with input $r(t)$ and outputs $\{c_i(t)\}_{i=1}^l$:

$$x_p(t) = \Phi_p\, x_p(t-1) + G_p\, r(t-1) + L_p\, u(t-1) \tag{3.54}$$

$$c(t) = H_p\, x_p(t-1), \tag{3.55}$$

where $c(t)$ is the ($l \times 1$) vector of the canceling signals at the microphones

$$c(t) = [c_1(t),\ c_2(t),\ \dots,\ c_l(t)]^T;$$

$x_p(t)$ is the state vector of the plant model; and $u(t)$ is the process noise representing the unmodelled dynamics in the plant model with

$$E\{u(t)u(\tau)^T\} = \delta(t-\tau)I.$$

Following the same steps as the ones for the single-microphone case, we now derive the state-space model for the overall system by combining (3.52-3.53) with (3.54-3.55):

$$x(t) = \Phi\, x(t-1) + G\, r(t-1) + L \begin{bmatrix} w(t-1) \\ u(t-1) \end{bmatrix} \tag{3.56}$$

$$m(t) = H\, x(t) + v(t), \tag{3.57}$$

78

where $x(t)$ is the state vector of the overall system

$$x(t) = \begin{bmatrix} x_d(t) \\ x_p(t) \end{bmatrix};$$

$m(t)$ is the $(l \times 1)$ vector of the microphone outputs

$$m(t) = \begin{bmatrix} m_1(t) \\ \vdots \\ m_l(t) \end{bmatrix};$$

Φ is the transition matrix of the overall system

$$\Phi = \begin{bmatrix} \Phi_d & 0 \\ 0 & \Phi_p \end{bmatrix};$$

matrix G is defined as

$$G = \begin{bmatrix} 0 \\ G_p \end{bmatrix};$$

matrix L is defined as

$$L = \begin{bmatrix} L_d & 0 \\ 0 & L_p \end{bmatrix};$$

and matrix H is defined as

$$H = [H_d \quad H_p].$$

The $(l \times 1)$ vector $v(t)$ is

$$v(t) = \begin{bmatrix} v_1(t) \\ \vdots \\ v_l(t) \end{bmatrix},$$

where $v_i(t)$ is the measurement noise at the i-th microphone and is assumed to be a white Gaussian process with variance σ_v^2. Furthermore, we shall assume that $v_i(t)$ and $v_j(t)$ are uncorrelated for $i \neq j$, and $v(t)$ is uncorrelated with $w(t)$ and $u(t)$.

Next we shall express the performance index $J(N)$ in terms of the state vector of the overall system. For concreteness, let $e(t)$ be the vector of the residual signals

$$e(t) = \begin{bmatrix} e_1(t) \\ \vdots \\ e_l(t) \end{bmatrix},$$

and let A be the $(l \times l)$ diagonal matrix

$$A = \begin{bmatrix} a_1 & & 0 \\ & \ddots & \\ 0 & & a_l \end{bmatrix}.$$

Recalling that

$$e(t) = H\, x(t),$$

we can rewrite $J(N)$ as

$$J(N) = E\left\{ \sum_{t=0}^{N-1} \left(x(t)^T H^T A H\, x(t) + \rho\, r^2(t) \right) + x(N)^T H^T A\, Hx(N) \right\}.$$

Since the performance measure $J(N)$ is the expected value of a quadratic form in the state and the control, the minimizing choice for $\{r(t) : t = 0, \ldots, N-1\}$ is again a linear feedback of the optimal state estimates [24]. The optimal state estimate can be computed using a Kalman filter, and the feedback gain can be computed by iterating an appropriate Riccati difference equation. The exact equations for the controller are obtained following steps identical to the ones followed in the single-microphone case.

3.4 Comparison of the Optimal Feedback Controller and the Optimal Feedforward Controller

We show in this Section that the optimal feedback controller can be thought of as a generalization of the optimal feedforward controller. To this end, let us consider the acoustic environment depicted in Fig. 3.10. We assume that the desired acoustic objective is to attenuate the pressure at Microphone 1 and that there is no feedback from the input $r(t)$ of the canceling loudspeaker $r(t)$ to the output of Microphone 2. Furthermore, we assume that the noise signals at the two microphones are jointly stationary. We will compare the effect of using a two-microphone feedback controller to the effect of using a single-channel feedforward controller for attenuating the acoustic pressure at Microphone 1. All signals in this Section are continuous-time signals.

For compatibility with the analysis done in Chapter 1, we assume in this Section that the output of each microphone is simply the value of the acoustic pressure at the location of the microphone, i.e. we ignore the microphone measurement noise.

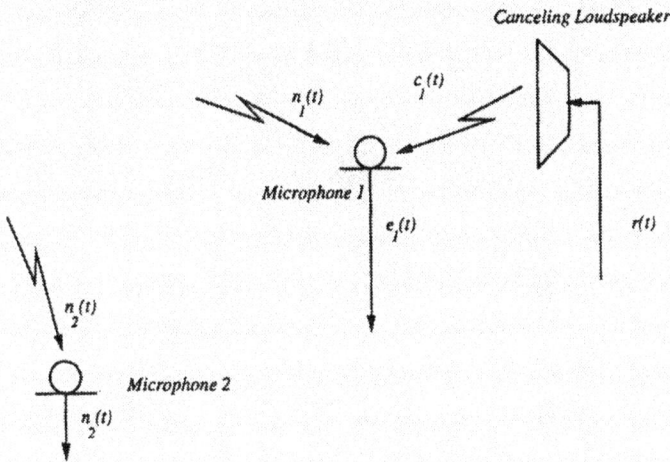

Figure 3.10: Noise cancellation environment with one canceling loudspeaker and two microphones.

In Fig. 3.11, the two-microphone feedback controller H^{fb} is used to attenuate the noise at Microphone 1. The block diagram for this feedback system is depicted in Fig. 3.13, where $G(s)$ is the transfer function from the input $r(t)$ of the canceling loudspeaker to the canceling field at the location of Microphone 1. The output of Microphone 1, $e_1(t)$, is the sum of the noise at this microphone, $n_1(t)$, and the canceling signal at this microphone $c_1(t)$. There is no feedback from the input of the canceling loudspeaker to the output of Microphone 2; hence, the output of Microphone 2 is simply the noise at this microphone

81

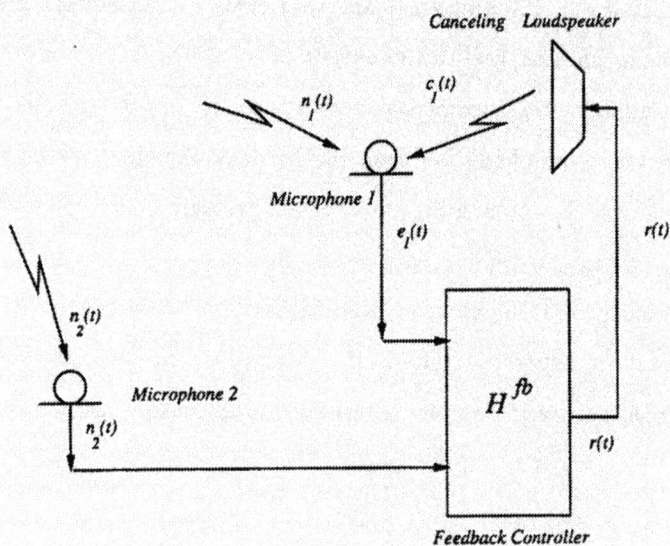

Figure 3.11: Two-microphone feedback controller.

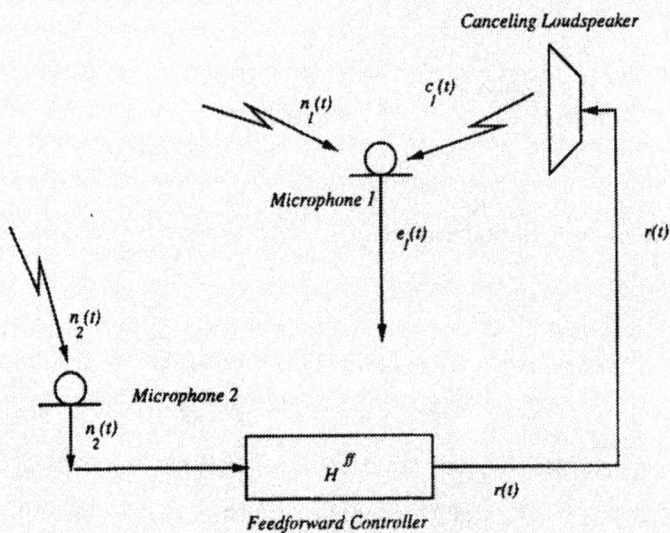

Figure 3.12: Single-channel feedforward controller.

82

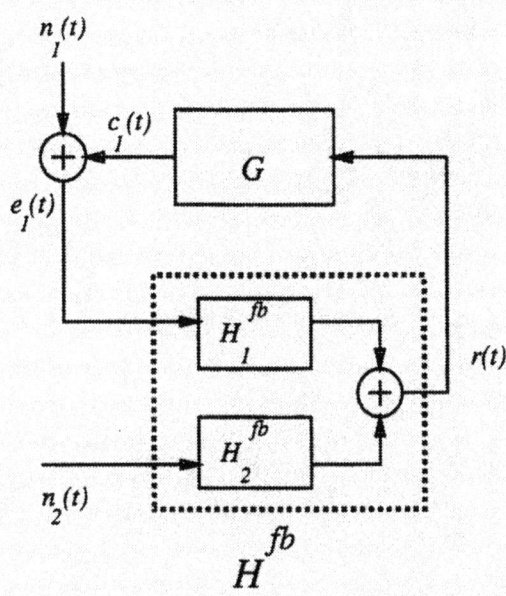

Figure 3.13: Block diagram of the two-microphone feedback ANC system.

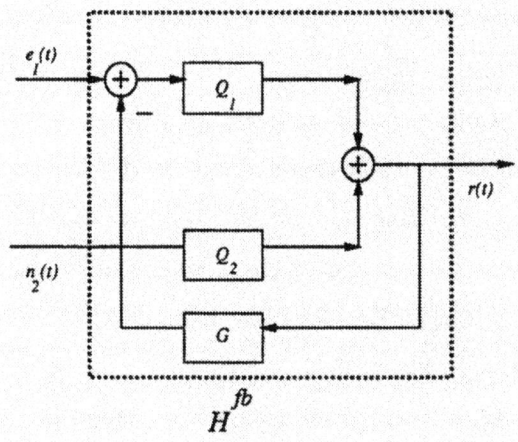

Figure 3.14: Restricted class of two-input/single-output controllers.

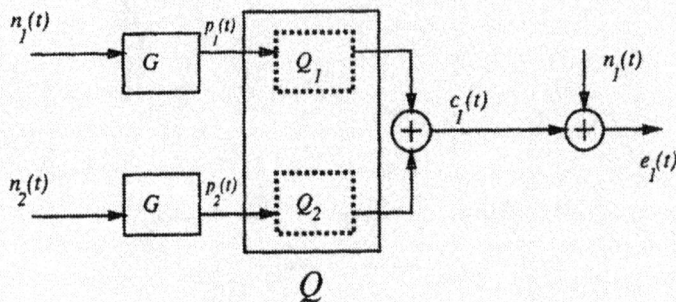

Figure 3.15: Alternate block diagram for the two-microphone feedback ANC system.

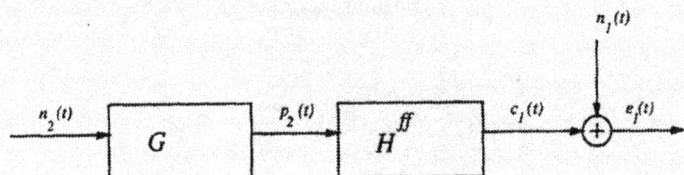

Figure 3.16: Block diagram for the single-channel feedforward ANC system.

$n_2(t)$. H_1^{fb} is the transfer function from $e_1(t)$ to $r(t)$, and H_2^{fb} is the transfer function from $n_2(t)$ to $r(t)$.

In Fig. 3.13, let us find the LTI controller H^{fb} that minimizes $E\{e_1^2(t)\}$. Our strategy for finding this controller is similar to the strategy we used to find the optimal single-channel feedback controller in Chapter 2, i.e. we first find the minimizing controller within a restricted class of controllers, and then show that no other controller can outperform this controller. We shall make the additional assumption that $G(s)$ is stable. This restricted class consists of all two-input/single-output LTI systems of the form depicted in Fig. 3.14, where $G(s)$ (the same as $G(s)$ in Fig. 3.13) is fixed, and $\{Q_1, Q_2\}$ can be varied. For a particular choice of $\{Q_1, Q_2\}$ in Fig. 3.14, the block diagram for the overall ANC system can be drawn as depicted in Fig. 3.15.

Looking at Fig. 3.15, we see that the minimizing choice for the canceling signal $c_1(t)$ is the LLSE of $-n_1(t)$ based on $\{p_1(\tau), p_2(\tau) : \tau \leq t\}$. This corresponds to choosing $\{Q_1, Q_2\}$ so that the resulting two-input/single-output system Q is the causal Wiener filter for estimating $-n_1(t)$ based on $\{p_1(\tau), p_2(\tau) : \tau \leq t\}$. An argument similar to the one given

84

in Chapter 2 can be used to show that no other controller can outperform this controller and that the optimal canceling signal at Microphone 1, $c_1(t)$ in Fig. 3.13, is unique.

In Fig. 3.12, a single-channel feedforward controller is used to attenuate the noise at Microphone 1. The block diagram corresponding to this system is depicted in Fig. 3.16, where $G(s)$ is the transfer function from the input of the canceling loudspeaker $r(t)$ to the value of the canceling field at the location of Microphone 1. The output of Microphone 1, $e_1(t)$, is the sum of the canceling signal at this microphone, $c_1(t)$, and the noise at this microphone $n_1(t)$. The reference signal for the feedforward controller is $n_2(t)$, the output of Microphone 2. The objective is to find the LTI controller H^{ff} that minimizes $E\{e_1^2(t)\}$.

According to the analysis presented in Chapter 2 for the optimal single-channel feedforward controller, the LTI system H^{ff} that minimizes $E\{e_1^2(t)\}$ in Fig. 3.16 is the causal Wiener filter for estimating $-n_1(t)$ based on $p_2(t)$, where $p_2(t) = n_2(t) * g(t)$. The resulting canceling signal in this case is the LLSE of $-n_1(t)$ based on $\{p_2(\tau) : \tau \leq t\}$.

Based on the above analysis, we can now compare the optimal feedback controller to the optimal feedforward controller. With the optimal two-microphone feedback controller, the canceling signal at Microphone 1 is the LLSE of $-n_1(t)$ based on $\{p_1(\tau), p_2(\tau) : \tau \leq t\}$. With the optimal single-channel feedforward controller, the canceling signal at Microphone 1 is the LLSE of $-n_1(t)$ based on $\{p_2(\tau) : \tau \leq t\}$. Therefore, with the optimal two-microphone feedback controller the residual signal at Microphone 1 is the estimation error in estimating $n_1(t)$ based on $p_1(t)$ and $p_2(t)$; however, with the optimal single-channel feedforward controller the residual signal at Microphone 1 is the estimation error in estimating $n_1(t)$ based on $p_2(t)$ alone. In this sense, the optimal feedforward controller does not use all the available measurements.

Let us note in passing that the optimal single-microphone feedback controller in this case, using Microphone 1 only, will result in a canceling signal at Microphone 1 that is the LLSE of $-n_1(t)$ based on $\{p_1(\tau) : \tau \leq t\}$ alone.

85

3.5 Summary

An optimal feedback controller for attenuating the acoustic pressure at a finite number of microphones was developed in this chapter. The results in this chapter were based on the following assumptions:

1. the unwanted noise signals at the microphones are modeled as the outputs of a known linear system driven by a white process.

2. the canceling signals at the microphones are modeled as the outputs of another known linear system driven by the control signal.

3. the residual signals at the microphones are modeled as the sum of the unwanted noise signals and the canceling signals.

4. the observed signals are modeled as the sum of the residual signals and white measurement noise, and this measurement noise is uncorrelated with all other processes involved.

Given a stochastic model for the unwanted noise at the microphones and a linear system model relating the canceling loudspeaker input of the outputs of the microphones, our formulation can be used to determine the minimum achievable value of the statistical performance criterion J which was defined in eq. (3.51). We also showed that our formulation can be modified to minimize the frequency weighted power of the residual signal. Moreover, our formulation can be used to find the specific feedback controller that achieves this minimum value.

The feedback controller developed in this chapter has a few advantages compared to the existing feedback controllers. Recall that the analog feedback controller developed by Wheeler [51] does not take advantage of the statistics of the noise. However, the feedback controller developed in this chapter takes full advantage of the statistical characteristics of the noise. The single-microphone feedback controller developed by Graupe [26] can not be used in cases for which the plant model contains one or more delays. However, the feedback controller developed in this chapter can be used in these cases. Furthermore, the feedback controller developed in this chapter can be used in the single-microphone as well as multiple-microphone configuration.

A comparison of the optimal feedback controller with the optimal feedforward controller was presented in Section 3.4. We showed that both controllers generate the canceling signal as the negative of the LLSE of the unwanted noise at the error microphone. However, the feedback controller estimates the unwanted noise based on all the available measurements, as opposed to the feedforward controller that estimates the unwanted noise based only on the reference signal. In this sense, the optimal feedback controller can be thought of as a generalization of the optimal feedforward controller.

The main drawback of the design procedure presented in this chapter is that the disturbance model and the plant model must be known before the optimal controller can be computed. The next chapter is devoted to the modeling of the unwanted noise based on the measurements of this noise made by the microphones.

Chapter 4

Modeling of the Unwanted Noise at a Single Microphone

4.1 Introduction

The design procedure of Chapter 3 is only useful if an algorithm for identifying the disturbance model and the plant model exists. The problem of identifying the disturbance model in the single-microphone case is considered in this chapter. We assume that the unwanted noise at the location of the microphone is a non-stationary autoregressive process and develop a recursive/adaptive procedure for estimating the parameters of this process based on the measurements made by the microphone.

An adaptive feedback ANC system can be obtained by combining the design procedure of Chapter 3 with an adaptive parameter estimation algorithm. Specifically, an adaptive algorithm can be used to continuously estimate the parameters of the disturbance model and the parameters of the plant model, and these parameter estimates can be used to continuously adjust the controller according to the design procedure developed in Chapter 3. This approach is commonly referred to as "indirect adaptive control" [58]. Hence, an optimal control algorithm such as that of Chapter 3 is an essential part of every indirect adaptive optimal control strategy for feedback active noise cancellation.

While it is eventually necessary to fully develop the algorithm by adaptively estimating the plant model and the disturbance model, we focus in this chapter on the simpler problem in which only the disturbance model is estimated. The algorithms presented in this chapter can be applied to the more general problem of identifying non-stationary autoregressive (AR) processes embedded in white noise.

4.2   Model Specifications

The single-microphone ANC system of interest and the corresponding block diagram are depicted in Fig. 4.1. System G relates the input $r(t)$ of the canceling loudspeaker to the value of the canceling field $c(t)$ at the location of the microphone. The microphone output $m(t)$ is the sum of $c(t)$ and the value of the noise field at the location of microphone $n(t)$ and the microphone measurement noise $v(t)$. The goal in this chapter is to develop a recursive/adaptive algorithm for modeling the noise $n(t)$ based on the measurements made by the microphone $m(t)$.

The overall strategy is based on the observation that if G is known exactly, then since $r(t)$ is known exactly, an estimate of the unwanted noise at the location of the microphone can be obtained by subtracting out the component of the microphone output due to the canceling source. Specifically, if we define $z(t)$ as

$$z(t) = n(t) + v(t), \tag{4.1}$$

it is clear that $z(t)$ can be computed from the measurements of $m(t)$ according to

$$z(t) = m(t) - G(\{r(t)\}). \tag{4.2}$$

Hence, if G is known exactly, the problem of identifying the disturbance model based on $m(t)$ reduces to the problem of identifying the disturbance model based on $z(t)$.

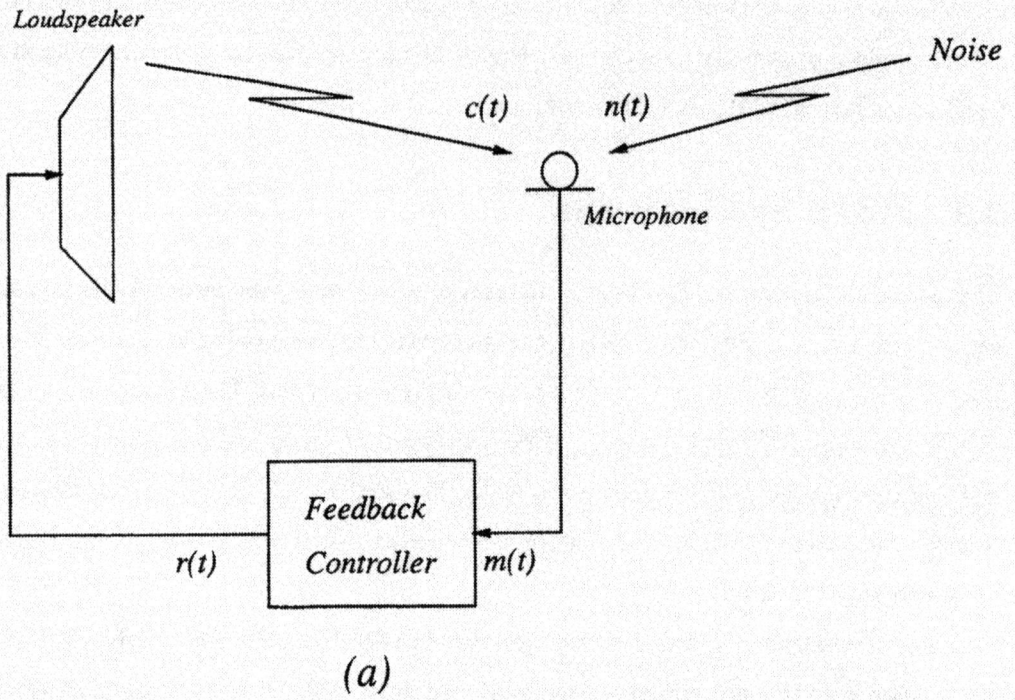

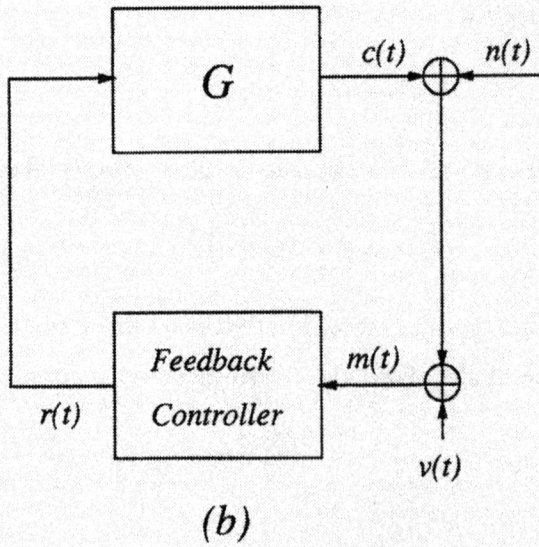

Figure 4.1: Single-microphone active noise cancellation system.

4.3   Single-Microphone Recursive/Adaptive Identification Algorithm

In the reminder of this chapter we focus on modeling of the unwanted noise $n(t)$ based on the observations of

$$z(t) = n(t) + v(t), \qquad (4.3)$$

where the microphone measurement noise $v(t)$ is assumed to be a zero-mean white Gaussian process with average power σ_v^2. Furthermore, it is assumed that $n(t)$ is an autoregressive process, i.e.

$$n(t) = - \sum_{k=1}^{p} \alpha_k n(t - k) + w(t), \qquad (4.4)$$

where $w(t)$ is a zero-mean white Gaussian process with average power of σ_w^2, and $w(t)$ is independent of $v(t)$. A more detailed discussion of the the algorithms developed in this chapter can be found in [55].

Using the state-space description of equations (4.3) and (4.4), we will develop two algorithms for estimating the parameters of the disturbance model ($\alpha_1, \ldots, \alpha_p$ and σ_w^2). We shall assume that the average power of the measurement noise, σ_v^2, is known. Equations (4.3) and (4.4) can be expressed in state-space form as [34]:

$$x(t) = \Phi_d x(t - 1) + L_d w(t) \qquad (4.5)$$

$$z(t) = H_d x(t) + v(t) \qquad (4.6)$$

where $x(t)$ is the $(p + 1) \times 1$ state vector defined by:

$$x(t) = [n(t), \; n(t - 1), \; \ldots, \; n(t - p)]^{\mathsf{T}}; \qquad (4.7)$$

Φ_d is the $(p+1) \times (p+1)$ transition matrix:

$$\Phi_d = \begin{bmatrix} -\alpha_1 & -\alpha_2 & \cdots & -\alpha_p & 0 \\ 1 & \ddots & 0 & \cdots & \cdots & 0 \\ 0 & \ddots & \ddots & & & \vdots \\ \vdots & \ddots & \ddots & \ddots & & \vdots \\ \vdots & & \ddots & \ddots & \ddots & \vdots \\ 0 & & \cdots & 0 & 1 & 0 \end{bmatrix} ; \tag{4.8}$$

L_d is the $(p+1) \times 1$ unit vector:

$$L_d = [1 \ 0 \ \ldots \ 0]^\top; \tag{4.9}$$

and H_d is the $(p+1) \times 1$ unit vector:

$$H_d = [1 \ 0 \ \ldots \ 0]^\top. \tag{4.10}$$

The parameters of the disturbance model satisfy the Yule-Walker equation:

$$R \begin{bmatrix} 1 \\ \alpha \end{bmatrix} = \begin{bmatrix} \sigma_w^2 \\ o \end{bmatrix} \tag{4.11}$$

where o is the $p \times 1$ vector of zeros; α is the $p \times 1$ vector of the AR parameters:

$$\alpha = [\alpha_1, \alpha_2, \ldots, \alpha_p]^\top; \tag{4.12}$$

and R is the $(p+1) \times (p+1)$ correlation matrix:

$$R = E\{x(t)x^\top(t)\}, \tag{4.13}$$

with $x(t)$ the state vector defined in eq. (4.7).

If we assume that the parameters of the disturbance model (α and σ_w^2) are precisely known, then an estimate of R can be computed using the optimal estimates of the state vector. To this end, let us denote by

$$\hat{x}(t|t) = E\{x(t)|z(1), \ldots, z(t)\} \tag{4.14}$$

92

the optimal state estimate based on data up to time "t", and by

$$P(t|t) = E\{[\hat{x}(t|t) - x(t)][\hat{x}(t|t) - x(t)]^\top | z(1), \ldots, z(t)\} \qquad (4.15)$$

the associated error covariance matrix. In accordance with (4.13), an estimate of R can be computed by replacing the expectation operator in eq. (4.13) with the weighted averaging:

$$\bar{R}(t) = \frac{1}{\sum_{\tau=1}^{t} \lambda^{t-\tau}} \sum_{\tau=1}^{t} \lambda^{t-\tau} \widehat{x(\tau)x^\top(\tau)} \qquad (4.16)$$

where

$$\widehat{x(t)x^\top(t)} \overset{\text{def}}{=} \hat{x}(t|t)\hat{x}^\top(t|t) + P(t|t) \qquad (4.17)$$

$$= E\{x(t)x^\top(t)|z(1), \ldots, z(t)\}. \qquad (4.18)$$

Then using the standard Kalman filter formulation $\hat{x}(t|t)$ and $P(t|t)$ in eq. (4.17) can be computed sequentially in time, in two stages, as follows[1]:

Propagation Equations:

$$\hat{x}(t|t-1) = \Phi_d \hat{x}(t-1|t-1) \qquad (4.19)$$

$$P(t|t-1) = \Phi_d P(t-1|t-1)\Phi_d^\top + \sigma_w^2 L_d L_d^\top \qquad (4.20)$$

Updating Equations:

$$\hat{x}(t|t) = \hat{x}(t|t-1) + k(t)[z(t) - H_d \hat{x}(t|t-1)] \qquad (4.21)$$

$$P(t|t) = [I - k(t)H_d]P(t|t-1), \qquad (4.22)$$

where $k(t)$ is the Kalman gain given by

$$k(t) = \frac{1}{H_d^\top P(t|t-1)H_d + \sigma_v^2} P(t|t-1)H_d. \qquad (4.23)$$

We would like to generate the parameters of the disturbance model in such a way that equation (4.11) is satisfied; however, this requires knowing matrix R, which in turn requires

---

[1]These equations can be simplified by exploiting the structure of Φ_d, L_d, and H_d, as was done in [56]; however, this is not essential to our development here.

knowing the exact parameters of the disturbance model. Instead, we propose generating these parameter estimates in such a way that the following equation is satisfied:

$$\hat{R}(t) \begin{bmatrix} 1 \\ \hat{\alpha}(t+1) \end{bmatrix} = \begin{bmatrix} \widehat{\sigma_w^2}(t+1) \\ \mathbf{0} \end{bmatrix}, \tag{4.24}$$

where $\hat{R}(t)$ is the estimate of R, obtained by performing the weighted averaging:

$$\hat{R}(t) = \frac{1}{\sum_{\tau=1}^{t} \lambda^{t-\tau}} \sum_{\tau=1}^{t} \lambda^{t-\tau} \widehat{x(\tau)x^{\mathsf{T}}(\tau)}, \tag{4.25}$$

where

$$\widehat{x(t)x^{\mathsf{T}}(t)} \overset{\text{def}}{=} \hat{x}(t|t)\hat{x}^{\mathsf{T}}(t|t) + P(t|t), \tag{4.26}$$

where $\hat{x}(t|t)$ and $P(t|t)$ are the estimate of the state and its covariance, computed using the Kalman filtering equations (4.19)-(4.23), where instead of α and σ_w^2 we use the most current estimates $\hat{\alpha}(t)$ and $\widehat{\sigma_w^2}(t)$, i.e.

Propagation Equations:

$$\hat{x}(t|t-1) = \hat{\Phi}_d(t)\hat{x}(t-1|t-1) \tag{4.27}$$

$$P(t|t-1) = \hat{\Phi}_d(t)P(t-1|t-1)\hat{\Phi}_d^{\mathsf{T}}(t) + \widehat{\sigma_w^2}(t)L_d L_d^{\mathsf{T}} \tag{4.28}$$

Updating Equations:

$$\hat{x}(t|t) = \hat{x}(t|t-1) + \hat{k}(t)[z(t) - H_d^{\mathsf{T}}\hat{x}(t|t-1)] \tag{4.29}$$

$$P(t|t) = [I - \hat{k}(t)H_d^{\mathsf{T}}]P(t|t-1), \tag{4.30}$$

where $\hat{\Phi}_d(t)$ is the matrix defined in eq. (4.8) computed at $\alpha = \hat{\alpha}(t)$, and $\hat{k}(t)$ is the vector defined in eq. (4.23) with $P(t|t-1)$ obtained from eq. (4.28).

Our goal now is to develop a sequential procedure for generating the parameter estimates in accordance with equation (4.24), i.e. the parameter estimates at time $t+1$, $\hat{\alpha}(t+1)$ and $\widehat{\sigma_w^2}(t+1)$, must satisfy equation (4.24). To obtain this sequential procedure for updating the parameter estimates, we define

$$\begin{array}{c} {}_{1} \updownarrow \\ {}_{p} \updownarrow \end{array} \begin{bmatrix} Q_{11}(t) & Q_{12}(t) \\ \hline Q_{21}(t) & Q_{22}(t) \end{bmatrix} = Q(t) = \sum_{\tau=1}^{t} \lambda^{t-\tau} \widehat{x(\tau)x^{\mathsf{T}}(\tau)} \tag{4.31}$$

$$\underset{1}{\longleftrightarrow} \quad \underset{p}{\longleftrightarrow}$$

94

$$= \widehat{\boldsymbol{x}(t)\boldsymbol{x}^\top(t)} + \lambda \boldsymbol{Q}(t-1). \qquad (4.32)$$

Using eq. (4.31) and eq. (4.25), we express eq. (4.24) as

$$\frac{1}{\sum_{\tau=1}^{t}\lambda^{t-\tau}}\left[\begin{array}{c|c} Q_{11}(t) & Q_{12}(t) \\ \hline Q_{21}(t) & Q_{22}(t) \end{array}\right]\left[\begin{array}{c} 1 \\ \hat{\alpha}(t+1) \end{array}\right] = \left[\begin{array}{c} \widehat{\sigma_w^2}(t+1) \\ \mathbf{o} \end{array}\right] \qquad (4.33)$$

which leads to the following update equations for the noise parameter estimates:

$$\hat{\alpha}(t+1) = -\boldsymbol{Q}_{22}^{-1}(t)\boldsymbol{Q}_{21}(t) \qquad (4.34)$$

$$\widehat{\sigma_w^2}(t+1) = \frac{1-\lambda}{1-\lambda^t}[\boldsymbol{Q}_{11}(t) + \boldsymbol{Q}_{12}(t)\hat{\alpha}(t+1)], \qquad (4.35)$$

where we have used the fact that $\frac{1}{\sum_{\tau=1}^{t}\lambda^{t-\tau}} = \frac{1-\lambda}{1-\lambda^t}$. We shall refer to eqs. (4.34)- (4.35) as the non-gradient algorithm.

At each time step the non-gradient algorithm first estimates the current state using the latest parameter estimates in eqs. (4.27) - (4.30), and then updates the parameter estimates using the state estimate just computed and its covariance in eqs. (4.34)- (4.35). The intermediate quantity $\boldsymbol{Q}(t)$ is computed recursively in time according to eq. (4.32).

The factor λ that appears in the cumulative averaging in eq. (4.31) is a number between 0 and 1. To maximize statistical stability we choose $\lambda = 1$. Choosing λ to be strictly smaller than 1 corresponds to exponential weighting that gives more weight to current data samples and results in an adaptive algorithm that is capable of tracking non-stationary changes in the structure of the data.

As an alternative to the parameter update equations (4.31)- (4.35), we may consider a gradient-search algorithm for solving the Yule-Walker equation (4.11). In this case, instead of the signal correlation matrix $\boldsymbol{R}$ we use its estimate given by eq. (4.25), and we proceed sequentially through the data using a stochastic gradient type procedure [43]. Replacing $\boldsymbol{R}$ in eq. (4.11) by its current estimate $\hat{\boldsymbol{R}}(t) = \frac{1-\lambda}{1-\lambda^t}\boldsymbol{Q}(t)$ where $\boldsymbol{Q}(t)$ is defined in eq. (4.31), we obtain:

$$\boldsymbol{Q}_{21}(t) + \boldsymbol{Q}_{22}(t)\boldsymbol{\alpha} = \mathbf{o} \qquad (4.36)$$

$$\sigma_w^2 - \frac{1-\lambda}{1-\lambda^t}[\boldsymbol{Q}_{11}(t) + \boldsymbol{Q}_{12}(t)\boldsymbol{\alpha}] = 0. \qquad (4.37)$$

From eqs. (4.36) and (4.37) and using an approach similar to that used in compound decision problems (e.g., see [25, 38, 42, 45, 54]), the following sequential update equations for the signal parameters are suggested:

$$\hat{\alpha}(t+1) = \hat{\alpha}(t) - \gamma_t[\mathbf{Q}_{21}(t) + \mathbf{Q}_{22}(t)\hat{\alpha}(t)] \tag{4.38}$$

$$\widehat{\sigma_w^2}(t+1) = \widehat{\sigma_w^2}(t) - \gamma_t\left[\widehat{\sigma_w^2}(t) - \frac{1-\lambda}{1-\lambda^t}[\mathbf{Q}_{21}(t) + \mathbf{Q}_{22}(t)\hat{\alpha}(t)]\right], \tag{4.39}$$

where $\mathbf{Q}_{ij}(t)$ $i,j = 1,2$ are computed recursively in t using eq. (4.32), and γ_t is the step-sizes of the stochastic gradient algorithm. We shall refer to eqs. (4.38)-(4.39) as the gradient algorithm.

The advantage of using the gradient algorithm specified by eq. (4.38) - (4.39) is that it does not require matrix inversion in contrast to the non-gradient algorithm specified by eqs. (4.34)-(4.35), and therefore the gradient algorithm is computationally simpler than the non-gradient algorithm.

4.4 Algorithm Performance

The results of applying the algorithms developed in the previous section to synthetically generated noise and to recorded aircraft noise are presented in this section.

4.4.1 Synthetic Noise

Two sets of experiments were performed in which the gradient algorithm of the previous section was applied to computer generated autoregressive time series. The first set of experiments was performed to show how close the estimated spectrum of the AR process matches the actual spectrum. The second set of experiments was performed to calculate the one-step-ahead prediction error that results from using the estimated parameters. This prediction error was then compared to the one-step-ahead prediction error that results from using the correct parameters. In all these experiments the model order and σ_v were assumed to be known exactly. The variance of the AR process ($E\{n^2(t)\}$ in (4.3)) was equal to 1 in these experiments, and the forgetting factor λ was set equal to 0.975.

The first set of experiments presented in this section illustrates how accurately the algorithms developed in this chapter can estimate the model for the unwanted noise that is needed to compute the optimal controller of Chapter 3. The design procedure of Chapter 3 relies solely on the second-order statistics of the unwanted noise and the measurement noise. Since the measurement noise is assumed to be white, we can equivalently say that the design procedure of Chapter 3 relies solely on the power spectrum of the unwanted noise and the variance of the measurement noise. Therefore, it is important to see how close the estimated power spectrum of the unwanted noise matches the actual power spectrum of the unwanted noise.

In Fig. 4.2(a) and Fig. 4.3(a), the doted curve is the spectrum of the actual underlying AR process, and the solid curve is the average estimated spectrum obtained by using the gradient algorithm. The solid curve is computed by averaging the estimated spectrum over 6000 consecutive points. The actual system parameters for these two figures are:

$$Fig.1 : A(z) = \frac{1}{1-1.53z^{-1}+0.6z^{-2}} \qquad \sigma_w = 0.234 \quad \sigma_v = 0.05$$
$$Fig.2 : A(z) = \frac{1}{1-1.8896z^{-1}+0.9025z^{-2}} \qquad \sigma_w = 0.0501 \quad \sigma_v = 0.05. \tag{4.40}$$

In other words, the AR process was generated as the output of $A(z)$ with a white noise input of variance σ_w^2.

In Fig. 4.2(b) and Fig. 4.3(b), the dotted horizontal line is the average one-step-ahead prediction error of a Kalman predictor that uses the correct system parameters. This is the optimum predictor, in the sense that no other predictor can achieve a lower average prediction error. The solid curve in these two figures is the average prediction error obtained by using the gradient algorithm.

4.4.2 Aircraft Noise

The results of applying both the gradient and the non-gradient algorithm to three types of aircraft noise are presented in this section. These two algorithms were used on the noise generated by a propeller aircraft, a helicopter, and a jet aircraft. Recordings of these three types of noise were made from a microphone placed inside a set of headphones approximately

97

two meters away from each aircraft. Note that these recordings correspond to $z(t)$ in eq. (4.3).

These experiments simulate an indirect adaptive feedback controller in which the plant is a known delay of M samples, i.e. $G(z) = z^{-M}$. In Fig. 4.1, with $G = z^{-M}$, the feedback controller that minimizes $E\{(n(t) + c(t))^2\}$ is the M-step-ahead predictor that generates the LLSE of $-n(t)$ based on $\{m(\tau) : \tau \leq t - M\}$, i.e. the optimal canceling signal $c(t)$ in this case is the LLSE of $-n(t)$ based on $\{m(\tau) : \tau \leq t - M\}$. The resulting residual signal $e(t)$ in this case is the corresponding M-step ahead prediction error, i.e.

$$e(t) = n(t) - E\{n(t)|m(\tau) : \tau \leq t - M\}. \tag{4.41}$$

The performance of the adaptive feedback controller in this case depends on how well $n(t)$ can be predicted based on $\{m(\tau) : \tau \leq t - M\}$, or equivalently based on $\{z(\tau) : \tau \leq t - M\}$. In short the residual signal in this case is simply the prediction error in predicting $n(t)$ based on $\{z(\tau) : \tau \leq t - M\}$ and based on the estimated parameters of the noise model.

Several simulations were performed to illustrate how well $n(t)$ can be predicted based on measurements of $\{z(\tau) : \tau \leq t - M\}$ and based on the estimated parameters of the model of the noise obtained from our gradient algorithm. In these simulations, the order of the AR model for the unwanted noise was assumed to be five, and σ_v was fixed at 5% of the standard deviation of $z(t)$. The algorithms were tried with the order of the AR model ranging from three to nine, and it was found that the performance of the algorithm improves very little by increasing the order beyond five. The variance of the microphone measurement noise, σ_v^2, can be estimated in the following way. The microphone is placed in a quiet environment with the ANC turned off so that the output of the microphone is approximately $v(t)$. An estimate of σ_v^2 can then be obtained as

$$\sigma_v^2 \approx \frac{1}{N} \sum_{t=1}^{N} v^2(t). \tag{4.42}$$

Figure 4.4 is a plot of the normalized prediction error versus the prediction time for the non-gradient algorithm. Similarly, Figure 4.5 is the plot of the normalized prediction error versus the prediction time for the gradient algorithm. In these two figures, the normalized prediction error is calculated as

$$\text{Normalized Prediction Error(dB)} = 10 \log_{10} \frac{E\{e^2(t)\}}{E\{z^2(t)\}}, \tag{4.43}$$

where $E\{z^2(t)\}$ is the average power of the original recorded noise, and $E\{e^2(t)\}$ is the average power of the prediction error.

An intuitive explanation for why the normalized prediction error of the propeller noise is less than the normalized prediction error of the helicopter noise and the normalized prediction error of the jet noise. The propeller noise is more narrow-band than the helicopter noise and the jet noise; therefore, the propeller noise results in the least prediction error.

Furthermore, it is evident from these set of experiments that the performance of the gradient algorithm is slightly better than the non-gradient algorithm. Furthermore, the gradient algorithm is much less computationally intensive than the non-gradient algorithm.

4.5 Summary

Two recursive/adaptive algorithms for modeling the unwanted noise at a single microphone were presented in this chapter. We argued that if the plant model is exactly known, these algorithms can be combined with the design procedure of Chapter 3 to obtain an adaptive feedback ANC system. Of course, it is eventually necessary to fully develop these algorithms so that the disturbance model and the plant model are both adaptively estimated.

We assumed that the unwanted noise was a non-stationary autoregressive process, and that the observed signal was the sum of this autoregressive process plus white measurement noise. We then developed two algorithms for estimating the parameters of this autoregressive process based on the observed data. The performance of these algorithms were studied by applying them to synthetically generated noise and recorded aircraft noise. It is worth noting that these algorithms can be applied to the more general problem of identifying non-stationary autoregressive processes embedded in white noise.

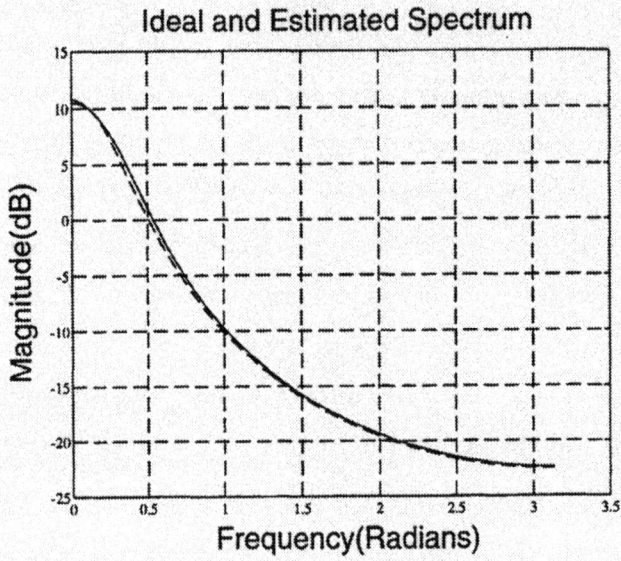

(a)

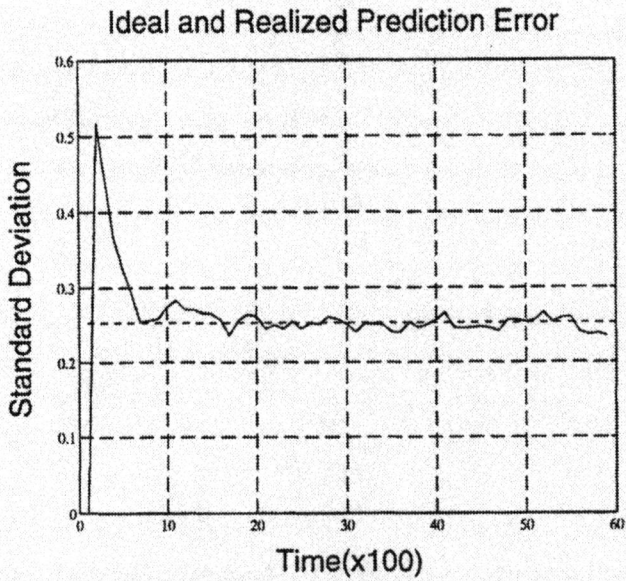

(b)

Figure 4.2: Performance of the gradient algorithm on synthetic noise.

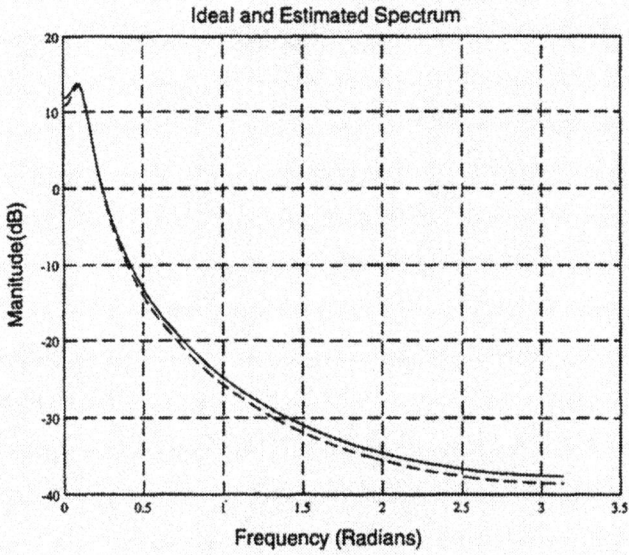

(a)

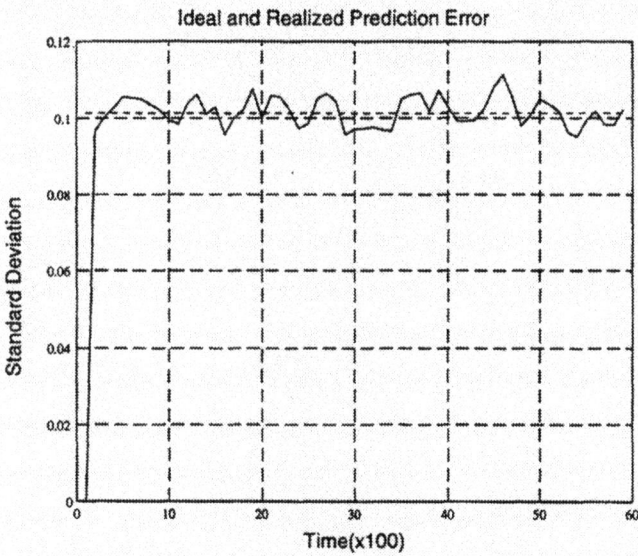

(b)

Figure 4.3: Performance of the gradient algorithm on synthetic noise.

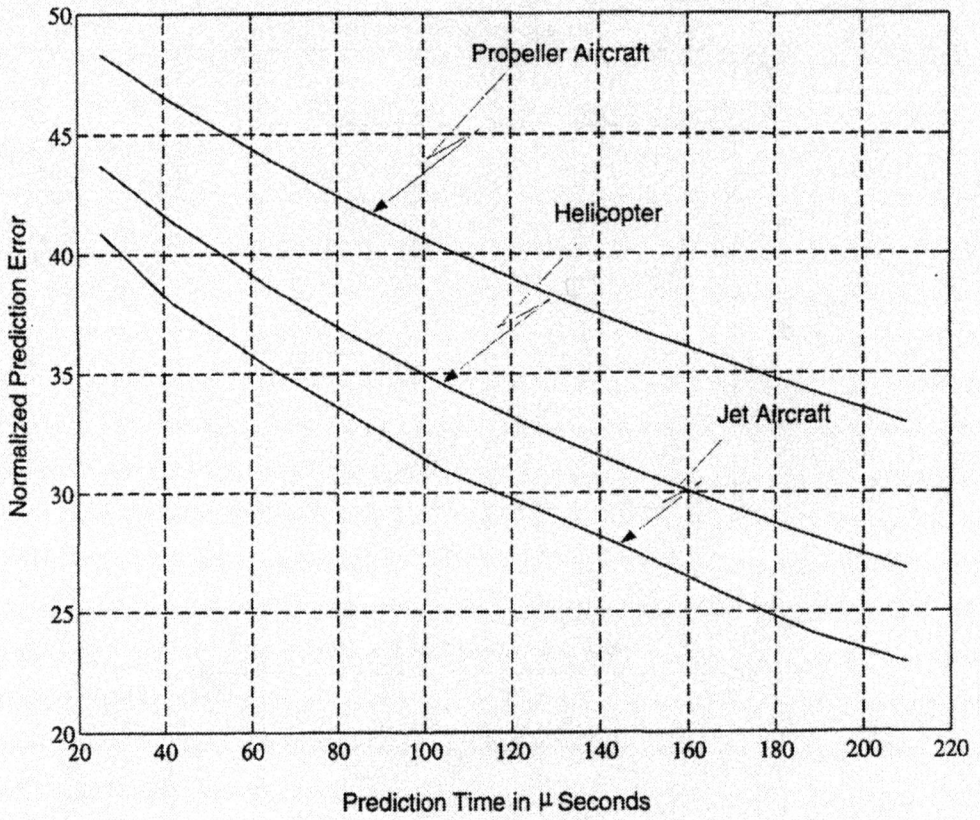

Figure 4.4: Performance of the non-gradient algorithm on recorded aircraft noise.

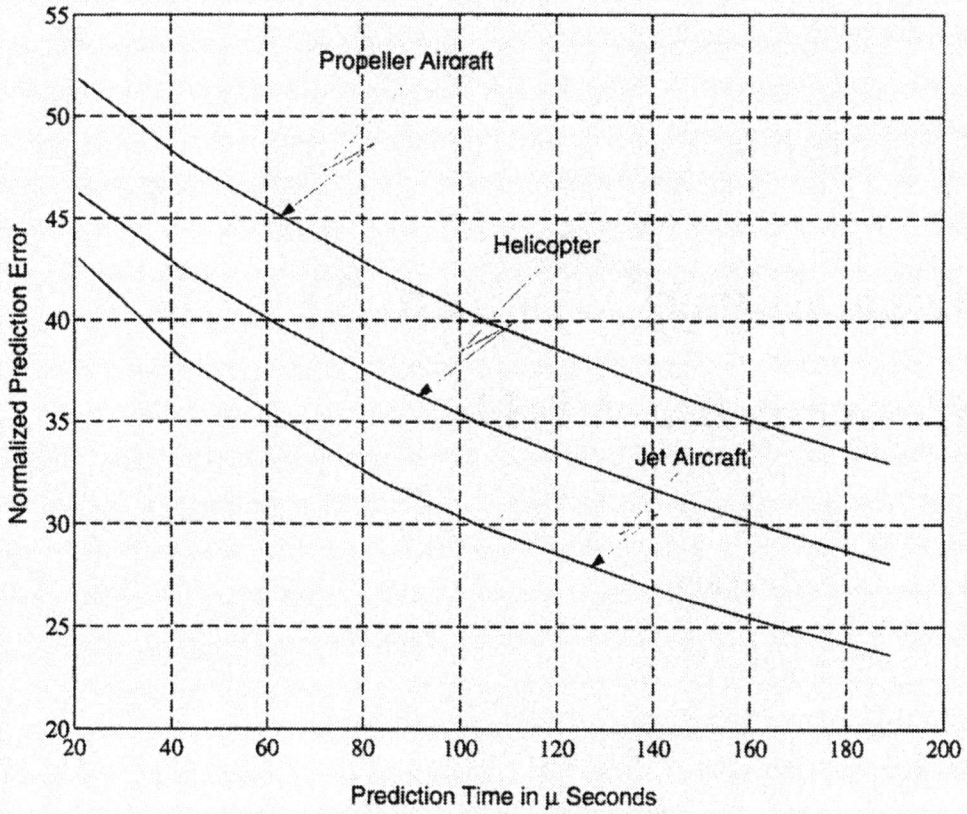

Figure 4.5: Performance of the gradient algorithm on recorded aircraft noise.

Chapter 5

Distributed Optimal Feedback Control for ANC

5.1 Introduction

Our treatment of the active noise cancellation (ANC) problem up to this point has been limited to pointwise active noise cancellation. In Chapter 3, we developed an optimal feedback controller for attenuating the acoustic pressure at the locations of a finite number of error microphones. Specifically, this controller minimizes a performance criterion involving the average power of the residual pressure field at the location of the error microphones and the average power of the control signal. Hence, this pointwise criterion depends only on the interaction between the primary field and the canceling field at those points where the microphones are located.

The distributed active noise cancellation problem is studied in this chapter. Specifically, we develop an optimal feedback controller for minimizing a performance criterion involving the total acoustic energy in an enclosure and the power expended by the control signal. To minimize this distributed performance criterion, the interaction between the primary field and the canceling field at every point in the enclosure needs to be considered.

Our approach is to formulate the distributed ANC problem as an infinite dimensional feedback control problem and use existing control theory to solve it. The deterministic formulation, which assumes that the entire state is observable, and the stochastic formulation, which estimates the state based on the available observations, are both considered. The main assumption is that the boundary surrounding the volume of interest is either purely reflecting or purely absorbing.

Our formulation of the distributed ANC problem in this chapter is remarkably similar to our formulation of the pointwise ANC problem in Chapter 3. In Chapter 3, the unwanted noise *signal* was modeled as the output of a linear system driven by white noise, and the canceling *signal* was modeled as the output of another linear system driven by the control *signal*. In this chapter, the unwanted noise *field* is modeled as the output of a linear system driven by white noise, and the canceling *field* is modeled as the output of yet another linear system driven by the control *signal*. In Chapter 3, the linear systems were finite dimensional and were characterized by differential equations. In this chapter, the linear systems are infinite dimensional and are characterized by partial differential equations. The optimal control in either case is a linear feedback of the state estimates. Furthermore, in both cases, the state estimates are computed using a Kalman filter, and the feedback gain is obtained from a Riccati equation.

Section 5.2 is essentially background material and terminology that is needed to understand our formulation of the distributed active noise cancellation problem in the deterministic case. The deterministic quadratic regulator problem is presented in Section 5.3. The acoustic system that we focus on and its governing equations are presented in Section 5.4. In Section 5.5, we derive an abstract differential equation representation for this acoustic system and show that the noise cancellation problem for this system is equivalent to a deterministic quadratic regulator problem.

The stochastic formulation of the distributed active noise cancellation problem is presented in Section 5.6. The necessary measure theoretic structure is introduced in Section 5.6.1, followed by the stochastic quadratic regulator problem in Section 5.6.2. In Section 5.6.3, the noise cancellation system that we focus on is modeled as a linear infinite-dimensional system driven by a white noise process and a control process. Furthermore, we show that this model can accommodate an observation process that is the output of a pressure microphone, and a control process that is the input to an ordinary loudspeaker. Finally, we show that the distributed active noise cancellation problem for the acoustic system that we focus on is equivalent to a stochastic quadratic regulator problem.

5.2 Preliminaries

This Section contains the terminology and background from functional analysis that is needed to understand our particular formulation of infinite-dimensional linear systems. General properties of semigroups are presented in Section 5.2.1, and abstract differential equations are presented in Section 5.2.2. Proofs for almost all the theorems in this Section can be found in [5].

5.2.1 General Properties of Semigroups

Let $T(t)$, $t \geq 0$, be a family of linear bounded transformations mapping a Hilbert space H into itself. $T(t)$ is said to be a "semigroup" if

(i) $T(0) = $ Identity

(ii) $T(t_1 + t_2) = T(t_2)T(t_1) = T(t_1)T(t_2)$.

The semigroup $T(t)$ is said to be strongly continuous if for each x in the Hilbert space H

(iii) $\|T(t)x - x\| \longrightarrow 0$ as $t \longrightarrow 0^+$.

Using (i)-(iii), one can show that $T(t)$ is bounded on bounded domains; that is, for each $L < \infty$, there exists an M such that

$$\underset{0 \leq t \leq L}{\text{Sup}} \ \|T(t)\| < M < \infty.$$

Moreover, we can find a dense subspace of H on which $T(t)$ is differentiable. That is, we can define the operator A (called the infinitesimal generator of $T(t)$) as

$$Ax = \lim_{\Delta \to 0^+} \frac{T(\Delta)x - x}{\Delta}$$

$$\mathcal{D}(A) = \left\{ x \in H \, | \, \lim_{\Delta \to 0^+} \frac{T(\Delta)x - x}{\Delta} \text{ exists} \right\},$$

106

where $\mathcal{D}(A)$ is the domain of A. Note that operator A is a mapping from $\mathcal{D}(A)$ into H. From the above definition of A it follows that $T(t)x$ is differentiable on $\mathcal{D}A$ for all $t > 0$ and

$$\frac{d}{dt}(T(t)x) = T(t)Ax = A(T(t)x) \quad \forall x \in \mathcal{D}(A).$$

This definition of A also implies that A is a closed linear operator and its domain is dense in H.

The relationship between the semigroup $T(t)$ and its infinitesimal generator A is similar to the relationship between the matrix exponential e^{At} and the matrix A where

$$\frac{d}{dt}e^{At}x = Ae^{At}x.$$

It will be seen later that this analogy can be carried much further. For example, $e^{At}x(0)$ is a solution to the system of ordinary differential equations $\frac{d}{dt}x(t) = Ax(t)$. Similarly, $T(t)x(0)$ is a solution to the abstract differential equation $\frac{d}{dt}x(t) = Ax(t)$, provided that A is the infinitesimal generator of the strongly continuous semigroup $T(t)$.

We now state two theorems regarding the necessary and sufficient conditions for a dissipative linear operator to be the infinitesimal generator of the strongly continuous semigroup $T(t)$. Recall that a closed linear operator A mapping a Hilbert space H into itself and with dense domain $\mathcal{D}(A)$ in H is said to be dissipative if

$$[Ax, x] + [x, Ax] \leq 0 \qquad \forall x \in \mathcal{D}(A). \tag{5.1}$$

We already have seen that the infinitesimal generator of a strongly continuous semigroup must necessarily be closed with dense domain in H. A closed linear operator that maps a Hilbert space into itself and has a dense domain will have a unique adjoint operator whose domain is also dense [33].

Theorem 1 Let A be a closed linear operator with dense domain in a Hilbert space H, and assume that A maps its domain into H. Suppose further that both A and A^*, the adjoint of A, are dissipative. Then A generates a strongly continuous semigroup $T(t)$ such that $\|T(t)\| \leq 1$ for all $t \geq 0$.

Theorem 2 Suppose A is a closed linear operator with dense domain $\mathcal{D}(A)$ in a Hilbert space H. Suppose further that A maps its domain into H, and

$$0 = [Ax, x] + [x, Ax] \quad \forall x \in \mathcal{D}(A)$$
$$0 = [A^*x, x] + [x, A^*x] \quad \forall x \in \mathcal{D}(A^*),$$

where $[.,.]$ is the inner product in H. Then, A generates a strongly continuous semigroup $T(t)$ such that $\|T(t)x\| = \|x\|$ $\forall x \in H$.

From the last two theorems, we see that the semigroup generated by a dissipative linear operator, call it $T(t)$, is a contraction, i.e.

$$\|T(t)x\| \leq \|x\| \quad \forall x \in H.$$

A contraction semigroup will obviously have the property that $\|T(t_2)x\| \leq \|T(t_1)x\|$ for all $t_1 \leq t_2$. This property is analogous to the following property for a matrix A: if $[Ax, x] + [x, Ax] \leq 0$ for all vectors x, then $\|e^{At_2}x\| \leq \|e^{At_1}x\|$ for all $t_1 \leq t_2$.

5.2.2 Abstract Differential Equations: Cauchy Problem

The initial value (Cauchy) problem for first order linear abstract differential equations is introduced in this Section. With the aid of semigroup theory, these abstract differential equations are put in a framework very similar to that of ordinary differential equations. A specific example is given to illustrate how a partial differential equation can be viewed as an abstract differential equation.

Let A be the infinitesimal generator of a strongly continuous semigroup $T(t)$ in a Hilbert space H. The initial value problem for the abstract differential equation:

$$\frac{dx(t)}{dt} = Ax(t) \qquad x(0) \text{ is given and is in} \mathcal{D}(A) \qquad (5.2)$$

is to find the function $x(t)$ with range in Hilbert space H that satisfies (5.2). Then, $x(t) = T(t)x(0)$ gives one solution, since $\frac{d}{dt}(T(t)x(0)) = A(T(t)x(0))$ and $T(0)x(0) = x(0)$. This is the only solution under the additional assumptions that:

1. $x(t) \in \mathcal{D}(A)$ $t \geq 0$

2. $x(t)$ is absolutely continuous for $t > 0$

3. $\|x(t) - x(0)\| \longrightarrow 0$ as $t \longrightarrow 0^+$.

From the above discussion, we see that in the case of abstract differential equations $T(t)$ plays a role analogous to the state transition matrix for ordinary differential equations.

Next we consider the nonhomogeneous equation

$$\frac{d}{dt}x(t) = Ax(t) + u(t) \qquad\qquad 0 \le t \le T \qquad\qquad (5.3)$$

with a given initial condition $x(0)$. By analogy with the finite dimensional case, we would expect that

$$x(t) = T(t)x(0) + \int_0^t T(t - s)u(s)ds$$

should be the unique solution in an appropriate sense. Depending on the smoothness of $u(t)$ and whether $x(0)$ is in $\mathcal{D}(A)$ or not, the sense in which (5.3) is satisfied will vary. The next two theorems are concerned with this issue.

Theorem 3 Suppose $x(0)$ is in $\mathcal{D}(A)$, and the function $u(t)$ with range in Hilbert space H is strongly continuously differentiable in the open interval $(0, T)$ with derivatives continuous in the closed interval $[0, T]$. Then

$$\frac{d}{dt}x(t) = Ax(t) + u(t) \qquad\qquad 0 < t < T$$

has a unique solution satisfying

1. $x(t) \in \mathcal{D}(A) \qquad\qquad t \ge 0$

2. $x(t)$ is absolutely continuous in (0,T)

3. $\|x(t) - x(0)\| \to 0$ as $t \to 0^+$,

and this solution is given by

$$x(t) = T(t)x(0) + \int_0^t T(t - s)u(s)ds.$$

If we wish to consider the case where neither the condition that $x(0)$ belongs to the domain of A, nor the condition of smoothness on the input $u(t)$ hold, we have to change the sense in which equation (5.3) holds.

109

Theorem 4 Suppose $u(.) \in \mathcal{L}_2((0,T);H)$, then there is one and only one function $x(t), 0 \leq t \leq T$, such that $[x(t), y]$ is absolutely continuous for each y in $\mathcal{D}(A^*)$, and

$$\frac{d}{dt}[x(t), y] = [x(t), A^*y] + [u(t), y] \quad a.e. \ 0 < t < T, \tag{5.4}$$

and such that for a given $x(0)$ in H,

$$\lim_{t \to 0^+} [x(t), y] = [x(0), y] \quad \forall y \in \mathcal{D}(A^*).$$

Moreover, this function is given by

$$x(t) = T(t)x(0) + \int_0^t T(t-s)u(s)ds.$$

The following example illustrates how a partial differential equation can be viewed as an abstract differential equation in an appropriate function space.

Example: Application to Partial Differential Equations

Consider the partial differential equation for one-dimensional heat transfer

$$\frac{\partial}{\partial t}f(t,x) = \frac{\partial^2}{\partial x^2}f(t,x) \quad t \geq 0 \ ; -\infty < x < \infty \quad ; f(0,x) \text{ is given.}$$

Let us show that this partial differential equation can be represented as an abstract differential equation of the type discussed in the previous Section. For this, we need to show that $\frac{\partial^2}{\partial x^2}$ is an infinitesimal generator of a strongly continuous semigroup. Note that it is not necessary to explicitly find the semigroup generated by $\frac{\partial^2}{\partial x^2}$; all that is needed to is to show that such a semigroup exists.

First we choose the underlying Hilbert space to be $H = \mathcal{L}_2(-\infty, \infty)$, the space of square integrable functions with respect to the Lebesgue measure. Next we take $A = \frac{\partial}{\partial x^2}$ as our linear operator with the following dense domain

$$\mathcal{D}(A) = \{f(x) \in H | f, f' \text{are absolutely continuous and } f', f'' \in H\}.$$

This means in particular that $f(-\infty) = f(\infty) = f'(\infty) = f'(-\infty) = 0$. Domain of A is dense in H, since the C^∞, the set of all infinitely differentiable functions in $L_2(-\infty, +\infty)$, is included in the domain of A, and C^∞ is dense in H as shown in [1].

110

Using integration by parts, we can show that operator A is self adjoint:

$$
\begin{aligned}
[g, Af] &= \int_{-\infty}^{\infty} g(x) \frac{\partial^2}{\partial x^2} f \, dx \\
&= -\int_{-\infty}^{\infty} \frac{\partial}{\partial x} g \frac{\partial}{\partial x} f \, dx \\
&= [Ag, f]. \qquad\qquad \forall f, g \in \mathcal{D}(A)
\end{aligned}
\qquad (5.5)
$$

Again using integration by parts, we see that A is dissipative since

$$
[f, Af] = \int_{-\infty}^{\infty} f \frac{\partial^2}{\partial x^2} f \, dx = -\int_{-\infty}^{\infty} \left(\frac{\partial f}{\partial x} \right) \left(\frac{\partial f}{\partial x} \right) dx \leq 0.
$$

It follows that operator A is closed, since it is self adjoint and A^* is closed by the virtue of being an adjoint operator [33]. A is a closed linear operator with dense domain in H, and both A and A^* are dissipative; therefore, by theorem 1, A generates a strongly continuous contraction semigroup $T(t)$. The original partial differential equation can now be represented as an abstract differential equation:

$$
\frac{d}{dt} f(t) = Af(t) \quad t \geq 0; \quad f(0) \in \mathcal{D}(A) \text{ is given,} \qquad (5.6)
$$

where

$$
\begin{aligned}
A &= \frac{\partial^2}{\partial x^2} \\
\mathcal{D}(A) &= \{f(x) \in H \,|\, f, f' \text{are absolutely continuous and } f', f'' \in H\}.
\end{aligned}
$$

Note that $f(t) = T(t)f(0)$, where $T(t)$ is the semigroup generated by A, is a solution to (5.6) since

$$
\frac{d}{dt}(T(t)f(0)) = A(T(t)f(0)),
$$

and $T(t)f(0)$ is in the domain of A for all values of t with $T(0)f(0) = f(0)$. In summary, we have shown that the operator $A = \frac{\partial^2}{\partial x^2}$ is the infinitesimal generator of a strongly continuous semigroup $T(t)$, and hence been able to represent the original partial differential equation as an abstract differential equation in the Hilbert space $\mathcal{L}_2(-\infty, +\infty)$.

Having discussed abstract differential equations, we will present the control problem for systems governed by these equations in the next Section.

5.3 Deterministic Linear Quadratic Regulator Problem

The linear quadratic regulator problem for systems governed by abstract differential equations is presented in this Section. We will see that under very mild conditions, a unique feedback solution to this problem exists.

Let A denote the infinitesimal generator of a strongly continuous semigroup of operators $T(t)$ over a Hilbert space H. The dynamic equation governing the problem is

$$\frac{d}{dt}x(t) = Ax(t) + Bu(t) \qquad 0 < t < T < \infty \tag{5.7}$$

$$x(0) \qquad \text{given,}$$

where B is a linear bounded transformation mapping a Hilbert space H_u into H, and $u(.)$ is an element of $\mathcal{W}_u = \mathcal{L}_2((0,T); H_u)$. It follows that $Bu(.)$ is an element of $\mathcal{L}_2((0,T); H)$. Hence, a unique solution to (5.7), in the sense of (5.4), exists and is given by

$$x(t) = T(t)x(0) + \int_0^t T(t-s)Bu(s)ds.$$

The optimal control problem is to find $u(t)$ in $\mathcal{W}_u$ so that the cost functional

$$J = \int_0^T [Rx(t), x(t)]_H dt + \int_0^T [u(t), u(t)]_{H_u} dt \tag{5.8}$$

is minimized, where R is a linear bounded nonnegative definite operator mapping H into H. The first integral in this functional is essentially what we want to minimize, and the second integral is the amount of power expended by the control input $u(t)$.

It is well known [5] that the above problem has a unique solution $u_o(t)$ which can be expressed in state feedback form as

$$u_o(t) = -B^* K(t)x_o(t), \tag{5.9}$$

where $K(t)$, for each t, is a linear bounded transformation from H into H and is absolutely continuous in t, $0 < t < T$. Furthermore, $K(t)$ satisfies the operator Riccati equation

$$\frac{d}{dt}[K(t)x, y] = -[Rx, y] - [K(t)x, Ay] - [K(t)Ax, y] \tag{5.10}$$

$$+[K(t)BB^* K(t)x, y]. \qquad K(t) = 0 \;\; \forall x, y \in \mathcal{D}(A)$$

Hence, if we can formulate the distributed active noise cancellation problem as a linear quadratic regulator problem, we will be able to use the above result to find the unique solution to the distributed active noise cancellation problem. Furthermore if such a solution exists, it will be expressible in feedback form.

The acoustic system that we focus on in this chapter and its governing equations are presented in the next Section. Our ultimate goal is to represent these equations as abstract differential equations of the type discussed in Section 5.2.2.

5.4 Dynamics of Two Deterministic Acoustic Systems

The governing equations of the acoustic systems that we focus on in this chapter are presented in this Section. The second order wave equation for pressure is broken up into two first order partial differential equations, so that it can be later represented as an abstract differential equation in Section 5.5.

Sound essentially involves a weak motion of a non-viscous fluid from an initial state of rest [16]. Particle velocity $\underline{v}(t, \underline{x})$ and pressure $p(t, \underline{x})$ are the two physical quantities whose values completely characterize any acoustic system, where the underline denotes a vector-valued quantity; t is the continuous-time variable; and $\underline{x}$ is the spatial variable. These two quantities are related according to the following partial differential equations:

$$\nabla p(t, \underline{x}) = -\rho_0 \frac{\partial \underline{v}(t, \underline{x})}{\partial t} + \underline{f}(t, \underline{x}) \tag{5.11}$$

$$\nabla . \underline{v}(t, \underline{x}) = -\frac{1}{c^2 \rho_0} \frac{\partial p(t, \underline{x})}{\partial t}, \tag{5.12}$$

where $\underline{f}(t, \underline{x})$ is an external force field; ρ_0 is the ambient density; c is the speed of sound; $\nabla . \underline{v}$ is the divergence of $\underline{v}$; and ∇P is the gradient of P. The first equation is obtained by applying Newton's law to a small control volume, and the second equation is obtained by applying the conservation of mass principle to this volume. The familiar wave equation for $p(t, \underline{x})$ can be derived from the above equations by differentiating eq. (5.12) with respect to time, taking the divergence of eq. (5.11), and eliminating $\nabla . (\frac{\partial \underline{v}(t, \underline{x})}{\partial t})$ to get:

$$\frac{1}{c^2} \frac{\partial^2 p(t, \underline{x})}{\partial t^2} = \nabla^2 p(t, \underline{x}) - \nabla . (\underline{f}(t, \underline{x})) \tag{5.13}$$

113

, where ∇^2 is the three dimensional Laplacian. Equations (5.11) and (5.12) are two first order partial differential equations equivalent to the second order partial differential equation (5.13). For us, it is more convenient to work with the following arrangement of (5.11)-(5.12):

$$\frac{\partial}{\partial t}\begin{bmatrix} \frac{P}{\rho_0} \\ c\,\underline{v}(t,\underline{x}) \end{bmatrix} = \begin{bmatrix} 0 & -c\,\nabla. \\ -c\,\nabla & 0 \end{bmatrix}\begin{bmatrix} \frac{P}{\rho_0} \\ c\,\underline{v}(t,\underline{x}) \end{bmatrix} + Bu(t), \tag{5.14}$$

where B is a bounded linear operator relating the control input $u(t)$ to the force field $\underline{f}(t,\underline{x})$ so that

$$Bu(t) = \begin{bmatrix} 0 \\ \frac{c}{\rho_0}\underline{f}(t,\underline{x}) \end{bmatrix}.$$

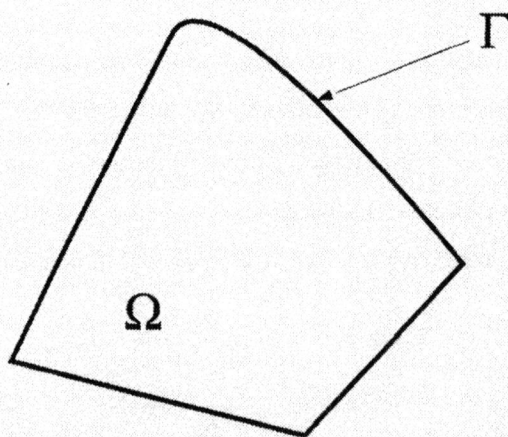

Figure 5.1: Enclosed region in R^3 with boundary Γ.

Next let us specialize the above equations to an enclosed region in R^3. We shall do this specialization for two types of boundary conditions. The first kind of boundary to be considered is a perfectly reflecting one, i.e. the normal component of velocity at the boundary surface is assumed to be zero. The second kind of boundary to be considered is an absorbing one, i.e. the ratio of the pressure at the boundary surface to the normal component of particle velocity at the boundary surface is assumed to be a known positive real number. This ratio is called the acoustic impedance of the the boundary; hence, in the second case we are assuming that the impedance of the boundary is purely resistive. The most general type of boundary condition is obtained when this ratio, in the frequency domain, is a complex valued function $H(\omega)$, where w is the frequency variable. The most general case is not considered in this chapter.

5.4.1 Purely Reflecting Boundaries

In Fig. 5.1, let Ω be a bounded open domain in R^3 with smooth boundary Γ. Furthermore, assume that eq. (5.14) is satisfied for all the interior points of Ω with

$$v_n(t, \underline{x}) = 0 \quad \underline{x} \in \Gamma, \tag{5.15}$$

where v_n is the normal component of velocity at the boundary. Finally, assume that the following initial conditions are given:

$$P(0, \underline{x}) = P_0(\underline{x}) \quad \text{and} \quad \underline{v}(0, \underline{x}) = \underline{v}_0(x).$$

The total acoustic energy of the system in Fig. 5.1 at time t, denoted by $E(t)$, is

$$E(t) = \int_\Omega \frac{1}{2c^2\rho_0} P^2(t, \underline{x}) dx + \int_\Omega \frac{\rho_0}{2} |\underline{v}(t, \underline{x})|^2 dx,$$

where the first integral is the total acoustic potential energy and the second integral is the total kinetic energy.

With the boundary condition eq. (5.15) and with no input $u(t)$, it can be easily shown [16] that $E(t)$ is constant as a function of t. This is consistent with our assumption that the boundary surfaces are lossless and that there is no input to the system; hence, there is no mechanism to increase or decrease the acoustic energy in this case.

5.4.2 Purely Absorbing Boundaries

Again in Fig. 5.1, let Ω be a bounded domain in R^3 with smooth boundary Γ. Furthermore, assume that eq. (5.14) is satisfied for all the interior points of Ω with the following boundary condition

$$p(t, \underline{x}) = +Z(\underline{x}) \, v_n(t, \underline{x}) \qquad \underline{x} \in \Gamma, \tag{5.16}$$

115

where v_n is the normal component of the velocity field at the boundary, and $Z(\underline{x})$ is a positive real number. Again, we assume that the following initial conditions are given:

$$P(0,\underline{x}) = P_0(\underline{x}) \text{ and } \underline{v}(0,\underline{x}) = \underline{v}_0(x).$$

The purpose of active noise cancellation is to choose the signal $u(t)$ in such a way that the resulting force field $\underline{f}(t,\underline{x})$ does negative work on the system and thereby reduces the acoustic energy of the system. Naturally, one would like to decrease $E(t)$ without expending too much power; hence, a reasonable performance criterion for an active noise cancellation system is the sum of the acoustic energy $E(t)$ and the power expended by the control input $u(t)$. We would like to formulate the active noise cancellation problem as an abstract quadratic regulator problem with the cost function equal to this sum. To this end, we first need to derive an abstract differential equation representation for the acoustic system of Figure 5.1.

5.5 Abstract Representation of Two Deterministic Acoustic Systems

The goal of this Section is to derive an abstract differential equation representation for the acoustic systems of Fig. 5.1 subject to the boundary condition specified by eq. (5.15) or specified by eq. (5.16). This derivation is achieved by showing that the differential operator

$$A = \begin{bmatrix} 0 & -c\,\nabla. \\ -c\,\nabla & 0 \end{bmatrix} \tag{5.17}$$

is an infinitesimal generator of an absolutely continuous semigroup in an appropriate Hilbert space. Once this abstract representation is established, it will be easy to show that the noise cancellation problem in Fig. 5.1 is equivalent to a quadratic regulator problem. The existence and uniqueness of the solution to the noise cancellation problem will then follow from the existence and uniqueness theorems for the solutions of the abstract quadratic regulator problem.

116

The first step toward establishing the abstract formulation is to choose the underlying Hilbert space and the domain of A in this space. For this, let us define

$$H^1(\Omega) = \left\{ f \in \mathcal{L}_2(\Omega) | \frac{\partial f}{\partial x_i} \text{ exists and } \frac{\partial f}{\partial x_i} \in \mathcal{L}_2(\Omega) \text{ for } i = 1,2,3 \right\},$$

where $\mathcal{L}_2(\Omega)$ is the class of square integrable functions with respect to the Lebesgue measure that are defined on Ω, and we further define

$$M = H^1(\Omega) \times H^1(\Omega) \times H^1(\Omega) \times H^1(\Omega). \tag{5.18}$$

The underlying Hilbert space H is

$$H = \mathcal{L}_2(\Omega) \times \mathcal{L}_2(\Omega) \times \mathcal{L}_2(\Omega) \times \mathcal{L}_2(\Omega), \tag{5.19}$$

with the following inner product

$$[e,h]_H = \sum_{i=1}^{4} [e_i, h_i]_{\mathcal{L}_2(\Omega)} \quad \forall e, h \in H.$$

5.5.1 Purely Reflecting Boundaries

In the case of purely reflecting boundaries, the linear operator A is defined as:

$$A \begin{bmatrix} p \\ \underline{v} \end{bmatrix} = \begin{bmatrix} -c \, \nabla.(\underline{v}) \\ -c \, \nabla(p) \end{bmatrix} \quad \forall \begin{bmatrix} p \\ \underline{v} \end{bmatrix} \in \mathcal{D}(A) \tag{5.20}$$

$$\mathcal{D}(A) = \left\{ \begin{bmatrix} p \\ \underline{v} \end{bmatrix} \in M \Big| \frac{\partial v(\underline{x})}{\partial n} = 0; \, \forall \underline{x} \in \Gamma \right\}, \tag{5.21}$$

where $\frac{\partial v(\underline{x})}{\partial n}$ is the normal component of the velocity field at the boundary. Using the divergence theorem in R^3, we can show that A^* equals the negative of A:

$$\left[\begin{bmatrix} g_1 \\ \underline{g}_2 \end{bmatrix}, A \begin{bmatrix} f_1 \\ \underline{f}_2 \end{bmatrix} \right]_H = \left[\begin{bmatrix} g_1 \\ \underline{g}_2 \end{bmatrix}, \begin{bmatrix} -c \, \nabla.(\underline{f}_2) \\ -c \, \nabla(f_1) \end{bmatrix} \right]_H \quad \forall \begin{bmatrix} f_1 \\ \underline{f}_2 \end{bmatrix} \in \mathcal{D}(A)$$

$$= -c \int_\Omega g_1(x) \nabla.(\underline{f}_2(x)) dx - c \int_\Omega [\underline{g}_2, \nabla(f_1)]_{R^3} dx \tag{5.22}$$

$$= c \int_\Omega [\underline{f}_2, \nabla (g_1)]_{R^3} dx - c \int_\Gamma \frac{\partial}{\partial n}(g_2 f_1) ds +$$

$$c \int_\Omega f_1 \nabla \cdot (\underline{g}_2) dx \qquad (5.23)$$

$$= \left[\left[\begin{bmatrix} c \nabla \cdot (\underline{g}_2) \\ c \nabla(g_1) \end{bmatrix}, \begin{bmatrix} f_1 \\ \underline{f}_2 \end{bmatrix} \right] \right]_H$$

$$-c \int_\Gamma \frac{\partial}{\partial n}(g_2 f_1) ds, \qquad (5.24)$$

where in transition from eq. (5.22) to eq. (5.23) we have used the divergence theorem in R^3 and the fact that on the boundary $\frac{\partial f_2}{\partial n} = 0$. This is essentially a three dimensional version of the integration by parts that was done to get eq. (5.5) in Section 5.2.2. The domain of operator A defined in eq. (5.21) is dense in H, since $C_0^\infty(\Omega)$, the set of all infinitely differentiable function with compact support in Ω, is dense in $\mathcal{L}_2(\Omega)$ as shown in [1].

We see that the surface integral in equation (5.24) must equal zero; hence, A^* is

$$A^* \begin{bmatrix} g_1 \\ \underline{g}_2 \end{bmatrix} = \begin{bmatrix} c \nabla \cdot (\underline{g}_2) \\ c \nabla(g_1) \end{bmatrix} = -A \begin{bmatrix} g_1 \\ \underline{g}_2 \end{bmatrix} \quad \forall \begin{bmatrix} g_1 \\ \underline{g}_2 \end{bmatrix} \in \mathcal{D}(A^*)$$

$$\mathcal{D}(A^*) = \left\{ \begin{bmatrix} g_1 \\ \underline{g}_2 \end{bmatrix} \in M \Big| \frac{\partial g_2(\underline{x})}{\partial n} = 0 \ \forall \underline{x} \in \Gamma \right\}.$$

Closedness of the operator A follows immediately since $A^* = -A$ and all adjoint operators are closed [33]. Next, using the divergence theorem, we show that A is dissipative:

$$\left[\left[A \begin{bmatrix} f_1 \\ \underline{f}_2 \end{bmatrix}, \begin{bmatrix} f_1 \\ \underline{f}_2 \end{bmatrix} \right] \right]_H = \left[\left[\begin{bmatrix} -c \nabla \cdot (\underline{f}_2) \\ -c\nabla(f_1) \end{bmatrix}, \begin{bmatrix} f_1 \\ \underline{f}_2 \end{bmatrix} \right] \right]_H$$

$$= -c \int_\Omega \nabla \cdot (\underline{f}_2) f_1 dx - c \int_\Omega [\nabla(f_1), \underline{f}_2]_{R^3} dx \qquad (5.25)$$

$$= c \int_\Omega [\underline{f}_2, \nabla(f_1)]_{R^3} dx - c \int_\Omega [\nabla(f_1), \underline{f}_2]_{R^3} dx \qquad (5.26)$$

$$= 0, \qquad (5.27)$$

where in transition from eq. (5.25) to eq. (5.26), we have used the divergence theorem and the fact that on the boundary $\frac{\partial f_2}{\partial n} = 0$.

118

Combining eq. (5.27) with $A^* = -A$, we get

$$0 = [Af, f] + [f, Af] \qquad \forall f \in \mathcal{D}(A) \tag{5.28}$$

$$0 = [A^*g, g] + [g, A^*g] = 0 \qquad \forall g \in \mathcal{D}(A^*). \tag{5.29}$$

We have therefore shown that A is a closed, linear and dissipative operator with dense domain in H. Moreover, A satisfies (5.28) and (5.29). Applying Theorem 2, it follows that A generates a strongly continuous semigroup $T(t)$ on H such that

$$\|T(t)f\|_H = \|f\|_H \ \forall f \in H.$$

From the last expression we see that $T(t)$ is norm preserving. This can be interpreted as saying that the acoustic energy of the system in Fig. 5.1 with the boundary condition specified by eq. (5.15) is constant, since the norm in H is proportional to the acoustic energy in the enclosure.

5.5.2 Purely Absorbing Boundaries

In the case of purely absorbing boundaries, the linear operator A is defined as:

$$A \begin{bmatrix} p \\ \underline{v} \end{bmatrix} = \begin{bmatrix} -c\,\nabla \cdot (\underline{v}) \\ -c\,\nabla(p) \end{bmatrix} \qquad \forall \begin{bmatrix} p \\ \underline{v} \end{bmatrix} \in \mathcal{D}(A) \tag{5.30}$$

$$\mathcal{D}(A) = \left\{ \begin{bmatrix} p \\ \underline{v} \end{bmatrix} \in M \mid p(\underline{x}) = Z(\underline{x})\, \frac{\partial \underline{v}(\underline{x})}{\partial n} \ \forall \underline{x} \in \Gamma \right\}. \tag{5.31}$$

Using the divergence theorem in R^3, we find A^* and its domain:

$$\left[\begin{bmatrix} g_1 \\ g_2 \end{bmatrix}, A \begin{bmatrix} f_1 \\ \underline{f_2} \end{bmatrix} \right]_H = \left[\begin{bmatrix} g_1 \\ g_2 \end{bmatrix}, \begin{bmatrix} -c\,\nabla \cdot (\underline{f_2}) \\ -c\,\nabla(f_1) \end{bmatrix} \right]_H \qquad \forall \begin{bmatrix} f_1 \\ \underline{f_2} \end{bmatrix} \in \mathcal{D}(A)$$

$$= -c \int_\Omega g_1(x)\nabla \cdot (\underline{f_2}(x))dx - c \int_\Omega [\underline{g_2}, \nabla\,(f_1)]_{R^3}dx \tag{5.32}$$

$$= c \int_\Omega [\underline{f_2}, \nabla\,(g_1)]_{R^3}dx - c \int_\Gamma \frac{\partial}{\partial n}(\underline{f_2}g_1)ds +$$

$$c \int_\Omega f_1 \nabla \cdot (\underline{g}_2) dx - c \int_\Gamma \frac{\partial}{\partial n} (\underline{g}_2 f_1) ds \qquad (5.33)$$

$$= \left[\left[\begin{bmatrix} c\, \nabla \cdot (\underline{g}_2) \\ c\, \nabla(g_1) \end{bmatrix}, \begin{bmatrix} f_1 \\ \underline{f}_2 \end{bmatrix} \right] \right]_H$$

$$- c \left\{ \int_\Gamma \frac{\partial}{\partial n} (\underline{f}_2 g_1) ds + \int_\Gamma \frac{\partial}{\partial n} (\underline{g}_2 f_1) ds \right\}, \qquad (5.34)$$

where in transition from eq. (5.32) to eq. (5.33) we have used the divergence theorem in R^3. The domain of operator A defined in eq. (5.31) is dense in H, since $C_0^\infty(\Omega)$ is included in the domain of A, and C_0^∞ is dense in H as shown in [1].

From eq. (5.34), we see that the expression in the brackets must equal zero. By equating this expression to zero and recalling eq. (5.16), we get the following expression for A^*:

$$A^* \begin{bmatrix} g_1 \\ \underline{g}_2 \end{bmatrix} = \begin{bmatrix} c\, \nabla \cdot (\underline{g}_2) \\ c\, \nabla(g_1) \end{bmatrix} = -A \begin{bmatrix} g_1 \\ \underline{g}_2 \end{bmatrix} \qquad \forall \begin{bmatrix} g_1 \\ \underline{g}_2 \end{bmatrix} \in \mathcal{D}(A^*)$$

$$\mathcal{D}(A^*) = \left\{ \begin{bmatrix} g_1 \\ \underline{g}_2 \end{bmatrix} \in M \,\middle|\, g_1(\underline{x}) = -Z(\underline{x}) \frac{\partial \underline{g}_2(\underline{x})}{\partial n} \ \forall \underline{x} \in \Gamma \right\}.$$

Using the above derivation for A^*, we see that operator A is the adjoint of the linear operator B defined as

$$B \begin{bmatrix} f \\ \underline{g} \end{bmatrix} = \begin{bmatrix} c\, \nabla \cdot (\underline{g}) \\ c\, \nabla(f) \end{bmatrix} \qquad \forall \begin{bmatrix} f \\ \underline{g} \end{bmatrix} \in \mathcal{D}(B) \qquad (5.35)$$

$$\mathcal{D}(B) = \left\{ \begin{bmatrix} f \\ \underline{g} \end{bmatrix} \in M \,\middle|\, f(\underline{x}) = -Z(\underline{x}) \frac{\partial \underline{g}(\underline{x})}{\partial n} \ \forall \underline{x} \in \Gamma \right\}. \qquad (5.36)$$

Operator A must then be closed, since the adjoint of B is closed. Recall that all adjoint operators are closed [33].

Next using the divergence theorem, we show that A is dissipative:

$$\left[\left[A \begin{bmatrix} f_1 \\ \underline{f}_2 \end{bmatrix}, \begin{bmatrix} f_1 \\ \underline{f}_2 \end{bmatrix} \right] \right]_H = \left[\left[\begin{bmatrix} -c\, \nabla \cdot (\underline{f}_2) \\ -c\nabla(f_1) \end{bmatrix}, \begin{bmatrix} f_1 \\ \underline{f}_2 \end{bmatrix} \right] \right]_H$$

120

$$= -c \int_\Omega \nabla \cdot (\underline{f_2}) f_1 \, dx - c \int_\Omega [\nabla(f_1), \underline{f_2}]_{R^3} \, dx \qquad (5.37)$$

$$= c \int_\Omega [\underline{f_2}, \nabla(f_1)]_{R^3} \, dx - c \int_\Gamma \frac{\partial}{\partial n} (\underline{f_2} f_1) \, ds$$

$$\quad -c \int_\Omega [\nabla(f_1), \underline{f_2}]_{R^3} \, dx \qquad (5.38)$$

$$= -c \int_\Gamma Z(\underline{x}) \frac{\partial}{\partial n} (\underline{f_2} f_1) \, ds$$

$$= -c \int_\Gamma Z(\underline{x}) \left(\frac{\partial}{\partial n} \underline{f_2} \right)^2 ds$$

$$\leq 0,$$

where in transition from eq. (5.37) to eq. (5.38) we have used the divergence theorem. To get the last expression, we have used the assumption that on the boundary $f_1(\underline{x}) = Z(\underline{x}) \frac{\partial}{\partial n} f_2(\underline{x})$.

The operator A^* is also dissipative, and this can be shown following steps very similar to the one used to show that A is dissipative.

We have shown that A is a closed, linear and dissipative operator with dense domain in H. Moreover, A^* has also been shown to be dissipative. Applying Theorem 1, it follows that A generates a strongly continuous semigroup $T(t)$ on H such that

$$\|T(t)f\|_H \leq \|f\|_H \quad \forall f \in H.$$

The last expression can be interpreted as saying that the mechanical energy of the system in Fig. 5.1 with the boundary condition specified by eq. (5.16) is decreasing. This is consistent with the fact that the boundaries in this case are assumed to be absorbing.

We have shown that the governing equation for the acoustic system of Fig. 5.1, with purely reflecting boundaries or purely absorbing boundaries, can be represented as an abstract differential equation of the form $\frac{d}{dt} x(t) = Ax(t) + Bu(t)$, where A is the infinitesimal generator of a strongly continuous semigroup.

We now write down the abstract differential equation representation for the system in Fig. 5.1, subject to the boundary condition specified by eq. (5.15) or by eq. (5.16). Assuming that $u(.)$ is an element of $\mathcal{W}_u = \mathcal{L}_2((0, T); H_u)$ and recalling that B is a bounded linear operator from the Hilbert space H_u into H, we arrive at the following abstract

121

representation of the acoustic system in Fig. 5.1:

$$\frac{d}{dt}x(t) = Ax(t) + Bu(t) \tag{5.39}$$

$$x(0) \quad \text{is given in } H,$$

where H is the underlying Hilbert space. Moreover, the operator A is defined by eq. (5.17), and its domain is defined by eq. (5.21) in the case of purely reflecting boundaries and by eq. (5.31) in the case of purely absorbing boundaries.

The distributed active noise cancellation problem can now be viewed as a linear quadratic regulator problem applied to the system governed by eq. (5.39). Specifically, the objective is to choose the input $u(t)$ such that the the cost functional

$$J = \int_0^T E(t)dt + \int_0^T [u(t), u(t)]_{H_u} dt \tag{5.40}$$

$$= \int_0^T [\frac{\rho_0}{2c^2}x(t), x(t)]_H dt + \int_0^T [u(t), u(t)]_{H_u} dt \tag{5.41}$$

is minimized, where $E(t)$ is the mechanical energy of the system at time t, and $[u(t), u(t)]$ is the power expended by the control input at time t. From eq. (5.9), we see that a unique minimizing control $u_o(t)$ in $\mathcal{W}_u$ exists and can be expressed in the state feedback form as

$$u_o(t) = -B^* K(t) x_o(t),$$

where the linear bounded operator $K(t)$ satisfies the operator Riccati equation

$$\frac{d}{dt}[K(t)x, y] = -[\frac{\rho_0}{2c^2}x, y] - [K(t)x, Ay] - [K(t)Ax, y]$$

$$+[K(t)BB^* K(t)x, y]. \qquad K(t) = 0 \ \forall x, y \in \mathcal{D}(A) \tag{5.42}$$

In summary, by formulating the distributed active noise cancellation problem as a linear quadratic regulator problem, we have been able to show that the distributed active noise cancellation problem has a unique solution. Furthermore, we have given an explicit expression for this solution as a linear feedback of the state of the system.

The main shortcoming of the above solution is that the control signal is generated based on the exact knowledge of the entire state of the system. In a practical system, the state is not known exactly and the control signal must be generated based on the observed data

122

alone (typically the observed data is the output of one or more microphones that are placed in the volume of interest). It turns out that the optimal control in this case is obtained by feeding back the estimate of the state. The resulting estimation and control problem for infinite-dimensional linear systems is considered in the next Section. We will show that the distributed active noise cancellation problem in which the control signal is based on the observed data alone can be viewed as a stochastic quadratic regulator problem. The noise cancellation problem can then be solved using the stochastic optimal control theory for distributed parameter systems.

5.6 Stochastic Optimal Control for ANC

In this Section, the stochastic analog of the linear quadratic control problem of Section 5.3 is presented. For this purpose a measure theoretic structure is introduced on the top of the topological structure. The measure theory is nonclassical in the sense that the measures are only finitely additive on the field of cylinder sets. The notion of a weak random variable suffices for the stochastic extension of the control problem of Section 5.3 with a crucial notion being that of "white noise". The development of the stochastic control problem presented here is entirely due to Balakrishnan [5].

5.6.1 Preliminaries

The measure theoretic background that is needed to understand our formulation of the stochastic quadratic regulator problem is presented in this Section. Measures on cylinder sets, weak random variable, "white noise" processes, and linear infinite-dimensional systems with white noise input are discussed.

Measure on Cylinder Sets

Cylinder sets and cylinder set measures are introduced in this Section. We start with $\mathcal{H}$ a separable real Hilbert space. Let us take any finite dimensional subspace $\mathcal{H}_m$ in $\mathcal{H}$. By a cylinder set we mean any set of the form

$$B + \mathcal{H}_m^c,$$

where B is a Borel subset of $\mathcal{H}_m$, and $\mathcal{H}_m^c$ is the orthogonal complement of $\mathcal{H}_m$. The Borel set B is then called the "base" of the cylinder, and $\mathcal{H}_m$ is called the "base" space. It can be shown that the class of cylinder sets C forms a field of sets. Furthermore, $\mathcal{H}$ itself is a cylinder set, and the Borel sets of $\mathcal{H}$ are the smallest σ-algebra that contains C.

Next we consider measures on the field of cylinder sets, these measures will be referred to as "cylinder set measures" and denoted by μ. Let Z be a cylinder set with base B in $\mathcal{H}_m$. Then by definition

$$\mu(Z) = v_m(B),$$

where $v_m(.)$ is a countably additive probability measure on the σ-algebra of Borel subsets of $\mathcal{H}_m$. Furthermore v_m must satisfy the following condition.

Compatibility Condition: In order that μ be well defined, it is necessary that if

$$Z = B + \mathcal{H}_m^c = (B + \mathcal{H}_p) + (\mathcal{H}_m + \mathcal{H}_p)^c,$$

where $\mathcal{H}_p$ is orthogonal to $\mathcal{H}_m$, then $v_m(B) = v_{m+p}(B + \mathcal{H}_p)$ where $v_{m+p}(.)$ is a countably additive probability measure on the Borel subsets $\mathcal{H}_m + \mathcal{H}_p$.

It is important to realize that a cylinder set measure is not necessarily countably additive on the field of cylinder sets. Consequently, a cylinder set measure can not in general be extended to the Borel sets of $\mathcal{H}$.

The following example of a cylinder set measure will be of prime importance in our development. Let R be any self adjoint nonnegative definite operator mapping $\mathcal{H}$ into $\mathcal{H}$. We define the measure on the Borel sets of any finite dimensional space $\mathcal{H}_m$ in the following way. Let $e_1, \ldots, e_m$ be an orthonormal basis in $\mathcal{H}_m$. The Borel sets in $\mathcal{H}_m$ are in one-to-one correspondence with the Borel sets in R^m via the transformation

$$x \longleftrightarrow \begin{bmatrix} [x, e_1] \\ \vdots \\ [x, e_m] \end{bmatrix},$$

124

where $x \in \mathcal{H}_m$ and $\begin{bmatrix} [x, e_1] \\ \vdots \\ [x, e_m] \end{bmatrix} \in R^m$. The measure on Borel sets of R^m is defined to be a

Gaussian measure with a moment matrix given by

$$r_{ij} = [Re_i, e_j] \qquad\qquad i, j = 1, \ldots, m.$$

The measure of any Borel set in $\mathcal{H}_m$ is now defined to be the Gauss measure of the corresponding Borel set in R^m. This measure is independent of the particular basis used. It can be verified that the compatibility condition is also satisfied. The resulting cylinder set measure is usually denoted by μ and is referred to as the measure induced by R. In the special case that R is the identity operator, the resulting cylinder set measure on $\mathcal{H}$ is referred to as the standard Gauss measure on $\mathcal{H}$.

Weak Random Variables

The usual definition of a random variable requires a probability triple $(\Omega, \mathcal{B}, p)$; where Ω is an abstract space, $\mathcal{B}$ is a σ-field of subsets of Ω, and p is a countably additive probability measure on $\mathcal{B}$. A random variable is then merely any function (usually with range in R^n) that is measurable with respect to $\mathcal{B}$. Typically, there is a field of subsets of Ω, call it F, which generates the σ-field $\mathcal{B}$. In this case, specifying p as a countably additive set function on F uniquely determines p on $\mathcal{B}$. In other words, p can be uniquely extended from F to $\mathcal{B}$, provided that p is countably additive on F.

A cylinder set measure μ can not necessarily be extended to the Borel subsets of $\mathcal{H}$, since a cylinder set measure is not necessarily countably additive on the field of cylinder sets. This implies that if B is a Borel set in $\mathcal{H}$, the measure of B is not generally defined based on the specification of μ as a cylinder set measure. This in turn implies that we can not adopt the usual definition of a random variable as any function $f(.)$ defined on $\mathcal{H}$ that is measurable with respect to the Borel sets of $\mathcal{H}$. The probability of the event that $f(.)$ is an element of a Borel set in the range space is simply not defined in general.

Hence, the notion of a "weak random variable" is introduced. Let $[H, C, \mu]$ denote the probability triple for which $\mathcal{H}$ is a real separable Hilbert space, C is the field of cylinder sets in $\mathcal{H}$, and μ is a cylinder set measure there on. We shall denote points in $\mathcal{H}$ by w in this context. Let $f(.)$ be a function from $\mathcal{H}$ into another Hilbert space $\mathcal{H}_r$. $f(.)$ is called a weak random variable if for each ϕ in $\mathcal{H}_r$, $\mu(.)$ is defined and countably additive on inverse images of one-dimensional Borel sets of the form

$$f^{-1}(E) = \{w | w \in \mathcal{H}, [f(w), \phi] \in E\},$$

where E is a one-dimensional Borel set, and ϕ is fixed element of $\mathcal{H}_r$. This definition essentially ensures that inverse images of cylinder sets in $\mathcal{H}_r$ are cylinder sets in $\mathcal{H}$.

The canonical example of a weak random variable is $f(w) = Lw$ where L is any bounded linear transformation from $\mathcal{H}$ into $\mathcal{H}_r$. Let μ_L denote the cylinder set measure induced by L on $\mathcal{H}_r$; that is, for any cylinder set C in $\mathcal{H}_r$, we define

$$\mu_L(C) = \mu(\{w | w \in \mathcal{H} \text{ and } Lw \in C\}) = \mu(L^{-1}(C)),$$

note that $L^{-1}(C)$ is a cylinder set in $\mathcal{H}$. If μ is induced by the self adjoint nonnegative operator R in $\mathcal{H}$, then μ_L is the cylinder set measure induced by LRL^* in $\mathcal{H}_r$.

Second Order Characterization

Second order characterization of weak random variables are reviewed in this Section. Notions of mean, covariance, and cross covariance for weak random variables are presented. Let ξ be a Hilbert space valued weak random variable. We shall say it has a finite first moment if

1. $E\{|[\xi, \phi]|\} < \infty$ for every ϕ in $\mathcal{H}$, and

2. $E\{[\xi, \phi]\}$ is continuous in ϕ,

where $E(.)$ denotes the expectation. Then there exists an element m in $\mathcal{H}$ such that $E\{[\xi, \phi]\} = [m, \phi]$. We shall use the notation $E(\xi) = m$ and call m the mean of ξ. Similarly, we shall say that ξ has a finite second moment if

126

1. $E\{[\xi, \phi]^2\} < \infty$ for every ϕ in $\mathcal{H}$, and

2. $E\{[\xi, \phi]^2\}$ is continuous in ϕ.

Let $\tilde{\xi} = \xi - m$, where $m = E(\xi)$, then $\tilde{\xi}$ also has a finite second moment, with first moment equal to zero. For any two elements x, y in $\mathcal{H}$, $E\{[\tilde{\xi}, x][\tilde{\xi}, y]\}$ is a continuous bilinear form over $\mathcal{H}$. Hence, there exists a bounded linear nonnegative self adjoint operator $R : \mathcal{H} \longrightarrow \mathcal{H}$, called the covariance of ξ, such that

$$E([\tilde{\xi}, x][\tilde{\xi}, y]) = [Rx, y].$$

In particular, if ξ is a Gaussian weak random variable so that

$$E\left(exp(i[\xi, \phi])\right) = exp\left(-\frac{[R\phi, \phi]}{2}\right)exp(i[m, \phi]),$$

it follows that ξ has a finite second moment, with covariance R and mean m.

The concept of cross covariance for weak random variables is introduced next. Let ξ, η be two Hilbert space valued weak random variables with finite second moment and zero first moments. Then $Q(x, y) = E([\xi, x][\eta, y])$ defines a continuous bilinear map, and hence there exists a bounded linear operator S such that $Q(x, y) = [Sx, y]$. We shall write

$$S = E(\xi\eta^*).$$

If $R_{\xi\xi}$ denotes the covariance of ξ, and $R_{\eta\eta}$ that of η, then it is consistent to use the following notation, analogous to the finite dimensional case,

$$R_{\xi\xi} = E(\xi\xi^*), \tag{5.43}$$

$$R_{\xi\eta} = E(\xi\eta^*), \tag{5.44}$$

$$R_{\eta\eta} = E(\eta\eta^*). \tag{5.45}$$

In the case that $R_{\xi\xi}$ is nuclear, we can define $E(\|\xi\|^2)$ as

$$E(\|\xi\|^2) = \sum_{k=1}^{\infty} E([\xi, \phi_k]^2) \tag{5.46}$$

$$= \sum_{k=1}^{\infty} [R_{\xi\xi}\phi_k, \phi_k] \tag{5.47}$$

$$= Tr.(R_{\xi\xi}), \tag{5.48}$$

where $\{\phi_k\}$ is an orthonormal basis for $\mathcal{H}$. Note that the nuclearity of $R_{\xi\xi}$ guarantees that $Tr.(R_{\xi\xi})$ exists and is finite.

White Noise

The crucial notion of white noise is introduced in this Section. Let $\mathcal{H}$ be a real separable Hilbert space, and let $\mathcal{W} = \mathcal{L}_2((0,T);\mathcal{H})$ with $0 < T \le +\infty$. Note that $\mathcal{W}$ is also a separable Hilbert space with the following inner product

$$[u,v]_\mathcal{W} = \int_0^T [u(t), v(t)]_\mathcal{H} dt. \qquad u, v \in \mathcal{W}$$

Let μ denote the standard Gauss measure on $\mathcal{W}$, and let w denote any element of $\mathcal{W}$. That is to say, μ is the measure induced by the identity operator on $\mathcal{W}$. Then, the "function space" process so obtained is called "white noise". Each $w(t)$ in $\mathcal{W}$ is called a white noise sample function. For any $u(t)$ in $\mathcal{W}$, $[w,u]_\mathcal{W}$ defines a zero mean Gaussian random variable with variance $[u,u]_\mathcal{W}$. It is natural that any physical observation such as $[w,u]_\mathcal{W}$ be a random variable and not merely a weak random variable.

There is a close analogy between the Karhunen-Loeve expansion of a white noise process in the finite-dimensional case and the projections of an infinite-dimensional white noise process along an orthonormal basis in $\mathcal{W}$. Recall that if $n(t)$ is a unit variance, scalar, Gaussian white noise process, and $\{\psi_k(t)\}_{k=1}^\infty$ is an orthonormal basis for $\mathcal{L}_2((-\infty, +\infty); R)$, then

$$\zeta_k(n) = [n(t), \psi_k(t)]_{\mathcal{L}_2}$$

are mutually independent zero mean, unit variance Gaussian random variables.

Similarly, if $\{\phi_k(t)\}_{k=1}^\infty$ is an orthonormal basis in $\mathcal{W}$ and $w(t)$ is a standard white noise process defined on $\mathcal{W}$, then

$$\eta_k(\omega) = [w(t), \phi_k(t)]_\mathcal{W}$$

are mutually independent zero mean, unit variance Gaussian random variables.

It is important to note that the definition of the white noise process restricts the sample space to be the class of square summable functions, since $\mathcal{W} = \mathcal{L}_2((0,T);\mathcal{H})$. This restriction becomes important when we consider systems governed by partial differential equations with white noise input.

128

Abstract Differential Equations with White Noise Input

By invoking the theory of semigroups we obtain a natural extension of linear systems with finite-dimensional state space to the infinite-dimensional case embracing in particular, systems described by partial differential equations with white noise input. Let A denote the infinitesimal generator of a strongly continuous semigroup $T(t)$ over the real separable Hilbert space $\mathcal{H}$. Let $W = \mathcal{L}_2((0,T);\mathcal{H})$ with T finite. Let $\mathcal{H}_n$, where subscript n stands for noise, be another real separable Hilbert space, and $W_n = \mathcal{L}_2((0,T);\mathcal{H}_n)$. Let $w(t)$ be a standard white noise process in W_n, and let B denote a bounded linear transformation mapping $\mathcal{H}_n$ into $\mathcal{H}$. Consider the abstract differential equation

$$\frac{d}{dt}x(t) = Ax(t) + Bw(t). \qquad 0 < t \leq T; \quad x(0)\text{given}$$

For each $w(t)$ in W_n, we have seen (Theorem 4, Section 5.2.2) that this equation has a unique solution such that $[x(t), y]$ is absolutely continuous for each y in $\mathcal{D}(A^*)$. Moreover, the solution is given by

$$x(t,w) = T(t)x(0) + \int_0^t T(t-s)Bw(s)ds,$$

where we have written the solution as $x(t,w)$, instead of $x(t)$, to emphasize the dependence of the solution on $w(t)$. Having discussed linear infinite-dimensional systems with white noise input, we will present the stochastic control problem for these systems next. Our goal is to ultimately formulate the active noise cancellation problem as such a control problem.

5.6.2 Stochastic Quadratic Regulator Problem

Stochastic analog of the quadratic regulator problem of Section 5.3 is presented in this Section. The main difference is that the control has to be based on the observed data only; hence, the control is forced to be "feedback" or "closed loop". Consider the state evolution equation

$$\frac{d}{dt}x(t,w) = Ax(t,w) + Bu(t,w) + Fw(t) \qquad 0 < t \leq T; \quad x(0) = 0, \qquad (5.49)$$

where A is the infinitesimal generator of a strongly continuous semigroup $T(t)$ over the Hilbert space $\mathcal{H}$, $w(t)$ is a standard white noise process in $W_n = \mathcal{L}_2((0,T);\mathcal{H}_n)$, $\mathcal{H}_n$ is a

129

separable real Hilbert space, and F is a linear bounded transformation mapping $\mathcal{H}_n$ into $\mathcal{H}$. The control process is $u(t,w)$ where $u(.,w)$ is in $\mathcal{W}_u = \mathcal{L}_2((0,T); \mathcal{H}_u)$, $\mathcal{H}_u$ is a separable Hilbert space, and we assume that B is a Hilbert-Schmidt operator mapping $\mathcal{H}_u$ into $\mathcal{H}$. The observation process is defined by

$$y(t,w) = Cx(t,w) + Gw(t) \qquad 0 < t < T, \tag{5.50}$$

where C is assumed to be a linear and bounded mapping from $\mathcal{H}$ into $\mathcal{H}_o$, where $\mathcal{H}_o$ is a real separable Hilbert space. Finally G maps $\mathcal{H}_n$ into $\mathcal{H}_o$, with

$$GG^* = \text{Identity} \tag{5.51}$$

and

$$FG^* = 0. \tag{5.52}$$

Note that eq. (5.51) implies that $Gw(t)$ is a white noise process, and eq. (5.52) implies that the observation noise, $Gw(t)$, is independent of the process noise $Fw(t)$.

The control is required to be such that

$$E(u(t,w)u(t,w)^*)$$

is nuclear, and further be determined for each "t" in terms of the observations up to time "t" by a linear transformation; specifically,

$$u(t,w) = \int_0^t W(t,s)y(s,w)ds \qquad 0 < t < T,$$

where for each $s \leq t$, $W(t,s)$ is a linear bounded mapping from $\mathcal{H}_o$ into $\mathcal{H}_u$; furthermore, $W(t,s)$ is strongly continuous in $0 \leq s \leq t \leq T$.

The optimal control problem is to find the control $u(t,w)$ that minimizes the cost functional

$$q(u) = \int_0^T Tr.E((Qx(t,w))((Qx(t,w))^*)dt + \int_0^T Tr.E(u(t,w)u(t,w)^*)dt, \tag{5.53}$$

where Q is Hilbert-Schmidt. We are only interested in the controls for which

$$\int_0^T Tr.E((u(t,w)u(t,w)^*))dt < \infty.$$

130

Utilizing the "separation principle", we can separate the filtering problem and the control problem. More precisely, the optimal control $u_o(t, w)$ is unique and can be expressed in the form

$$u_o(t, w) = -B^* K(t)\hat{x}(t, w), \tag{5.54}$$

where the operator $K(t)$ satisfies the following operator Riccati equation

$$\frac{d}{dt}[K(t)x, y] = -[Q^*Qx, y] - [K(t)x, Ay] - [K(t)Ax, y] \tag{5.55}$$
$$+[K(t)BB^*K(t)x, y] \qquad K(T) = 0, \ \forall x, y \in \mathcal{D}(A),$$

and

$$\hat{x}(t, w) = E\{x(t, w)|y(s, w) \ \ 0 \le s \le t\}.$$

Furthermore, $\hat{x}(t, w)$ is calculated using a Kalman filter, i.e.

$$\frac{d}{dt}\hat{x}(t, w) = A\hat{x}(t, w) + P_f(t)C^*(y(t, w) - C\hat{x}(t, w)) + Bu(t, w) \qquad \hat{x}(0, w) = 0,$$

where

$$P_f(t) = E\{(x(t, w) - \hat{x}(t, w))(x(t, w) - \hat{x}(t, w))^*\}$$

and $P_f(t)$ satisfies

$$\frac{d}{dt}[P_f(t)x, y] = [P_f(t)A^*x, y] + [P_f(t)x, A^*y] + [FF^*x, y] \tag{5.56}$$
$$-[P_f(t)C^*CP_f(t)x, y] \qquad P(0) = \Lambda \qquad \forall x, y \in \mathcal{D}(A^*).$$

Next we show the close relationship between the solution to the optimal control problem in the deterministic case and stochastic case. Recall from Section 5.3 that the deterministic optimal control problem is to find $u(t)$ that minimizes the cost functional

$$q_d(u) = \int_0^T [Rx(t), x(t)]dt + \int_0^T [u(t), u(t)]dt, \tag{5.57}$$

where R is a bounded nonnegative definite operator used as a weighting factor. Setting $R = Q^*Q$, we can rewrite $q_d(u)$ as

$$q_d(u) = \int_0^T \|Qx(t)\|^2 dt + \int_0^T \|u(t)\|^2 dt. \tag{5.58}$$

Next using the notation introduced in eq. (5.46), we can rewrite eq. (5.53) as

$$q(u) = \int_0^T E\{\|Qx(t,w)\|^2\}dt + \int_0^T E\{\|u(t,w)\|^2\}dt, \qquad (5.59)$$

since both $Qx(t,w)$ and $u(t,w)$ have nuclear covariance operators. Comparing eq. (5.58) to eq. (5.59), we see that the only difference between the two is that the integrands in the stochastic cost functional are the expected values of the integrands in the deterministic cost functional.

According to eqs. (5.9) and (5.54), the optimal control in the deterministic case is given by

$$u_o = -B^*K(t)x(t), \qquad (5.60)$$

and the optimal control in the stochastic case is given by

$$u_o = -B^*K(t)\hat{x}(t,w). \qquad (5.61)$$

Comparing eq. (5.10) to eq. (5.55), we see that $K(t)$ in eq. (5.60) is identical to $K(t)$ in eq. (5.61), provided that $R = Q^*Q$. In other words, the optimal control in the deterministic case is obtained by feeding back the state, $x(t)$, through a linear operator, and the optimal control in the stochastic case is obtained by feeding back the state estimate, $\hat{x}(t,w)$, through the same linear operator.

The results of this Section will be very important to us later, since these results imply that if we can formulate the noise cancellation problem as a stochastic control problem, we will be able to show that the noise cancellation problem has a unique solution. Furthermore if such a solution exists, it will be expressible in feedback form according to eq. (5.61).

The details of the acoustic system that we focus on and its governing equations are presented in the next Section. Our goal is to ultimately represent these equations in the format of eqs. (5.49)-(5.50) and use the results of this Section to find the closed-loop feedback controller that minimizes the total acoustic energy of this system.

5.6.3 Abstract Representation of a Stochastic Acoustic System

The acoustic system that we focus on is depicted in Fig. 5.2. In this figure, Ω is an open domain in R^3 with smooth boundary Γ, and for concreteness, we assume that Γ is perfectly

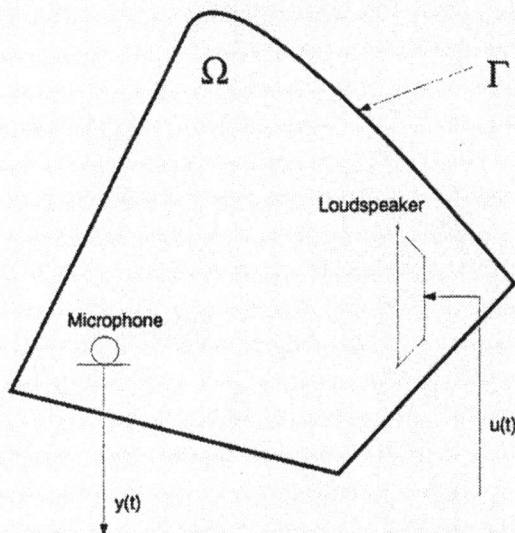

Figure 5.2: Bounded region in R^3 with loudspeaker and microphone.

reflecting. The control signal $u(t)$ drives the loudspeaker that generates the canceling field, and the output of the microphone $y(t)$ is the observed data. Loosely speaking, the goal of the noise cancellation system is to minimize the expected value of the acoustic energy in Ω, without expending too much power. The main restriction is that the control signal at time "t" must be generated based on the observed data up to time "t", i.e. based on $\{y(\tau): \ \tau \leq t\}$.

First, we develop an abstract model for the acoustic system of Fig. 5.2 as an infinite-dimensional linear system driven by a white noise process and a control process. Our approach is to model the acoustic field in Ω as the sum of two separate fields: the unwanted noise field and the canceling field. The unwanted noise field is modeled as the state of an infinite-dimensional linear system driven by white noise, and the canceling field is modeled as the state of another infinite-dimensional linear system driven by the control signal. These two fields are then added together to obtain the residual acoustic field that is present when the noise canceling system is on.

Having represented the acoustic system of Fig. 5.2 as an infinite-dimensional linear system, we use the results of the previous Section to find the closed-loop feedback controller

133

that minimizes the total acoustic energy in Ω.

Unwanted Noise Field

The unwanted noise field in Ω is modeled as the state, $x_d(t)$, of an infinite-dimensional linear system driven by white noise, i.e.

$$\frac{d}{dt}x_d(t) = A_d x_d(t) + F_d w(t), \tag{5.62}$$

where A_d is an infinitesimal generator of a strongly continuous semigroup $T_d(t)$ over the Hilbert space H, with H defined in eq. (5.19); $w(t)$ is a standard white noise process in $\mathcal{W}_n = \mathcal{L}_2((0,T); H)$; and F_d is a linear bounded transformation mapping H into H. The state vector $x_d(t)$ is composed of the unwanted pressure field and the unwanted particle velocity field in the enclosure, i.e.

$$x_d(t) = \begin{bmatrix} \frac{P_d(t,\underline{x})}{\rho_0} \\ \\ c\underline{v}_d(t,\underline{x}) \end{bmatrix}, \tag{5.63}$$

where $P_d(t,\underline{x})$ is the unwanted pressure field, and $\underline{v}_d(t,\underline{x})$ is the unwanted velocity field.

Equation (5.62) determines the second order statistics of the unwanted field, i.e. it determines $E(x_d(t,w))$ and $E(x_d(t,w)x_d^*(s,w))$.

Canceling Field

The canceling field is modeled as the state, $x_p(t)$, of another infinite-dimensional linear system driven by the control signal $u(t)$. The canceling field is generated by driving the canceling loudspeaker with the control input $u(t)$; hence, the exact nature of this infinite-dimensional system depends on the dynamics of the loudspeaker as well as the dynamics of the acoustic enclosure.

The simplest model for a loudspeaker is a collection of point sources of pressure [6]. Denoting the input to the loudspeaker by $u(t)$, we model the loudspeaker as "m" point sources, located at $\{\eta_1, \ldots, \eta_m\}$ respectively. Each one of these point sources has the strength $r(t)\delta(\eta - \eta_i)$, where $r(t)$ is related to the control signal $u(t)$ according to the

134

following finite-dimensional state-space equations:

$$\frac{d}{dt}x_l(t) \;=\; A_l x_l(t) + B_l u(t) \tag{5.64}$$

$$r(t) \;=\; C_l x_l(t), \tag{5.65}$$

where $x_l(t)$ is the $(n \times 1)$ state vector; A_l is the $(n \times n)$ state-transition matrix; B_l is an $(n \times 1)$ vector; and C_l is a $(1 \times n)$ row vector. Equations (5.64)-(5.65) model the dynamics of the loudspeaker.

In this case, the canceling pressure filed $p(t, \underline{x})$ satisfies the following wave equation

$$\frac{1}{c^2}\frac{\partial^2 p(t, \underline{x})}{\partial^2 t} = \nabla^2 p(t, \underline{x}) - r(t) \sum_{i=1}^{m} \delta(\eta - \eta_i), \tag{5.66}$$

where $r(t)$ is related to the control signal $u(t)$ according to eqs. (5.64)-(5.65), and $\delta(.)$ is a sufficiently smooth approximation to the delta function. We claim that equation (5.66) can be expressed in standard state-space form as:

$$\frac{d}{dt}x_p(t) = A_p x_p(t) + B_p r(t), \tag{5.67}$$

where the linear operator A_p is

$$A_p \begin{bmatrix} p \\ \underline{v} \end{bmatrix} = \begin{bmatrix} -c\nabla \cdot (\underline{v}) \\ -c\nabla(p) \end{bmatrix} \qquad \forall \begin{bmatrix} p \\ \underline{v} \end{bmatrix} \in \mathcal{D}(A_p) \tag{5.68}$$

$$\mathcal{D}(A_p) = \left\{ \begin{bmatrix} p \\ \underline{v} \end{bmatrix} \in M | \frac{\partial v(\underline{x})}{\partial \underline{n}} = 0; \; \forall \underline{x} \in \Gamma \right\}, \tag{5.69}$$

and B_p is a bounded linear operator from R^1 into H defined as

$$B_p z = \begin{bmatrix} 0 \\ \frac{c}{\rho_0} \sum_1^m h_i(\eta)z \\ \frac{c}{\rho_0} \sum_1^m h_i(\eta)z \\ \frac{c}{\rho_0} \sum_1^m h_i(\eta)z \end{bmatrix} \qquad \forall z \in R^1, \tag{5.70}$$

135

where $\frac{c}{\rho_0}$ is the ratio of the speed of sound to the ambient density, and

$$\nabla \cdot \left(\left[\begin{array}{c} h_i(\eta) \\ h_i(\eta) \\ h_i(\eta) \end{array} \right] \right) \approx \delta(\eta - \eta_i) \quad \text{with} \quad h_i(\eta) \in H^1(\Omega). \tag{5.71}$$

This claim can be verified by substituting eq. (5.70) in eq. (5.13) and observing that the resulting pressure field satisfies eq. (5.66).

The existence of a function $h_i(\eta)$ that satisfies eq. (5.71) can be argued in the following way. Let the three-dimensional Fourier transform of $h_i(\eta)$ be denoted by $FT\{h_i(\eta)\}$ and observe that eq. (5.71) implies

$$FT\{h_i(\eta)\} \approx \frac{e^{(-j[w,\eta_i]_{R^3})}}{j(\omega_1 + w_2 + w_3)},$$

where $w = [w_1 \ w_2 \ w_3]^T$ is the frequency variable. Therefore, $\hat{h}_i(\eta) = FT^{-1}\{\frac{e^{(-j[w,\eta_i]_{R^3})}}{j(\omega_1 + w_2 + w_3)}\}$ is the desired function except that $\hat{h}_i(\eta)$ is in $\mathcal{L}_2(\Omega)$ and not necessarily in $H^1(\Omega)$. However, $\hat{h}_i(\eta)$ can be arbitrary closely approximated by functions in $H^1(\Omega)$, since $H^1(\Omega)$ is dense in $\mathcal{L}_2(\Omega)$ [39].

By combining the finite-dimensional state-space equations (5.64)-(5.65) with the infinite-dimensional state-space equation (5.67), we will obtain a single infinite-dimensional state-space representation for the system relating the control input $u(t)$ to the canceling field $x_p(t)$. Note that the combined system is obtained by cascading the finite-dimensional system with the infinite-dimensional system. Specifically, the combined system can be expressed in state-space form as:

$$\frac{d}{dt}x_{pl}(t) = A_{pl}x_{pl}(t) + B_{pl}u(t), \tag{5.72}$$

where $x_{pl}(t)$ is the state vector of the combined system

$$x_{pl}(t) = \left[\begin{array}{c} x_p(t) \\ x_l(t) \end{array} \right]; \tag{5.73}$$

136

the linear operator A_{pl} is defined as

$$A_{pl} \begin{bmatrix} x_p \\ x_l \end{bmatrix} = \begin{bmatrix} A_p & BC_l \\ 0 & A_l \end{bmatrix} \begin{bmatrix} x_p \\ x_l \end{bmatrix} \tag{5.74}$$

$$\mathcal{D}(A_{pl}) = \left\{ \begin{bmatrix} x_p \\ x_l \end{bmatrix} \mid x_p \in \mathcal{D}(A_p) \text{ and } x_l \in R^n \right\} ; \tag{5.75}$$

and B_{pl} is a bounded linear operator from R^1 into $(H \times R^n)$ defined as

$$B_{pl} y = \begin{bmatrix} 0 \\ B_l y \end{bmatrix} \qquad \forall y \in R^1. \tag{5.76}$$

It remains to be shown that operator A_{pl} is an infinitesimal generator of a strongly continuous semigroup in an appropriate Hilbert space. The appropriate Hilbert space in this case is

$$H' = H \times R^n, \tag{5.77}$$

where H is defined in equation (5.19), and the inner product on H' is defined as

$$\left[\left[\begin{bmatrix} x_p \\ x_l \end{bmatrix}, \begin{bmatrix} y_p \\ y_l \end{bmatrix} \right] \right]_{H'} = [x_p, y_p]_H + [x_l, y_l]_{R^n}. \tag{5.78}$$

Let $T_l(t)$ denote the strongly continuous semigroup generated by A_l, and let $T_p(t)$ denote the strongly continuous semigroup generated by A_p. We claim that operator A_{pl} is the infinitesimal generator of the strongly continuous semigroup $T_{pl}(t)$ defined as

$$T_{pl}(t) \begin{bmatrix} x_p \\ x_l \end{bmatrix} = \begin{bmatrix} T_p(t)x_p + \int_0^t T_p(t-s)BC_l T_l(s)x_l ds \\ T_l(t)x_l \end{bmatrix} \qquad \forall \begin{bmatrix} x_p \\ x_l \end{bmatrix} \in H'. \tag{5.79}$$

Note that for a fixed t and a fixed x_l, the integral in (5.79) is a Riemann integral, since $\|T_p(t-s)BC_l T_l(s)\|$ is bounded and $T_p(t-s)BC_l T_l(s)x_l$ is a continuous function of s.

First, we show that $T_{pl}(t)$ is a strongly continuous semigroup on Hilbert space H', and then we show that operator A_{pl} is the infinitesimal generator of this semigroup.

Clearly, $T_{pl}(0)$ is the identity operator, and it is easy to show that $\lim_{t \to 0+} \|T_{pl}(t)x - x\| = 0$ for all x in H'. To show that $T_{pl}(t_1 + t_2) = T_{pl}(t_2)T_{pl}(t_1) = T_{pl}(t_1)T_{pl}(t_2)$, we observe that for any $\begin{bmatrix} x_p \\ x_l \end{bmatrix}$ in H' we have

$$T_{pl}(t_2)\left(T_{pl}(t_1)\begin{bmatrix} x_p \\ x_l \end{bmatrix}\right) = T_{pl}(t_2)\left(\begin{bmatrix} T_p(t_1)x_p + \int_0^{t_1} T_p(t_1 - s)BC_l T_l(s)x_l ds \\ T_l(t_1)x_l \end{bmatrix}\right) \tag{5.80}$$

$$= \begin{bmatrix} T_p(t_1 + t_2)x_p + T_p(t_2)\left(\int_0^{t_1} T_p(t_1 - s)BC_l T_l(s)x_l ds\right) + \\ \int_0^{t_2} T_p(t_2 - s)BC_l T_l(s)T_l(t_1)x_l ds \\ \\ T_l(t_1 + t_2)x_l \end{bmatrix} \tag{5.81}$$

$$= \begin{bmatrix} T_p(t_1 + t_2)x_p + \int_0^{t_1} T_p(t_1 + t_2 - s)BC_l T_l(s)x_l ds + \\ \int_0^{t_2} T_p(t_2 - s)BC_l T_l(s)T_l(t_1)x_l ds \\ \\ T_l(t_1 + t_2)x_l \end{bmatrix} \tag{5.82}$$

$$= \begin{bmatrix} T_p(t_1 + t_2 x_p + \int_0^{t_1 + t_2} T_p(t_1 + t_2 - s)BC_l T_l(s)x_l ds \\ T_l(t_1 + t_2)x_l \end{bmatrix} \tag{5.83}$$

$$= T_{pl}(t_1 + t_2)\begin{bmatrix} x_p \\ x_l \end{bmatrix}, \tag{5.84}$$

where in transition from eq. (5.81) to eq. (5.82), we have used the fact that

$$T_p(t_2)\left(\int_0^T T_p(s)x ds\right) = \int_0^T T_p(t_2)T_p(s)x ds.$$

This fact can be proved by noting that the $T_p(t_2)$ is a linear and continuous operator and the integral is a Riemann integral.

Similarly, we can show that $T_{pl}(t_1)T_{pl}(t_2) = T_{pl}(t_1 + t_2)$. Hence, $T_{pl}(t)$ is a strongly continuous semigroup in the Hilbert space H'. We now show that operator A_{pl} is the infinitesimal generator of the semigroup $T_{pl}(t)$. To this end, take any $\begin{bmatrix} x_p \\ x_l \end{bmatrix}$ in H' and observe that

$$\lim_{\Delta \to 0^+} \frac{T_{pl}(\Delta)\begin{bmatrix} x_p \\ x_l \end{bmatrix} - \begin{bmatrix} x_p \\ x_l \end{bmatrix}}{\Delta} = \lim_{\Delta \to 0^+} \begin{bmatrix} \frac{T_p(\Delta)x_p - x_p}{\Delta} + \frac{1}{\Delta}\int_o^\Delta T_p(t-s)BC_lT_l(s)x_l ds \\ \frac{T_l(\Delta)x_l - x_l}{\Delta} \end{bmatrix}$$

$$= \begin{bmatrix} Ax_p + BC_l x_l \\ A_l x_l \end{bmatrix} \tag{5.85}$$

$$= \begin{bmatrix} A & BC_l \\ 0 & A_l \end{bmatrix}\begin{bmatrix} x_p \\ x_l \end{bmatrix} \tag{5.86}$$

$$= A_{pl}\begin{bmatrix} x_p \\ x_l \end{bmatrix}. \tag{5.87}$$

Combined State-Space Representation for the Unwanted Noise Field and the Canceling Field

A single state representation for the acoustic field in Ω is derived in this Section by combining the state-space representation for the unwanted noise field and the state-space representation for the canceling field.

We propose the following state-space representation for the acoustic field in Ω:

$$\frac{d}{dt}x_c(t) = A_c x_c(t) + B_c(t)u(t) + F_c w(t), \tag{5.88}$$

where $x_c(t)$ is the state vector

$$x_c(t) = \begin{bmatrix} x_d(t) \\ x_{pl}(t) \end{bmatrix}; \tag{5.89}$$

139

A_c is a linear operator on Hilbert space $H_c = H \times H'$

$$A_c \begin{bmatrix} x_d \\ x_{pl} \end{bmatrix} = \begin{bmatrix} A_d & 0 \\ 0 & A_{pl} \end{bmatrix} \begin{bmatrix} x_d \\ x_{pl} \end{bmatrix} \tag{5.90}$$

$$\mathcal{D}(A_c) = \left\{ \begin{bmatrix} x_d \\ x_{pl} \end{bmatrix} \mid x_d \in \mathcal{D}(A_d) \text{ and } x_{pl} \in \mathcal{D}(A_{pl}) \right\}; \tag{5.91}$$

B_c is a bounded linear operator from R_1 into H_c

$$B_c y = \begin{bmatrix} 0 \\ B_{pl} y \end{bmatrix} \qquad \forall y \in R^1, \tag{5.92}$$

F_c is a bounded linear operator from H into H_c

$$F_c z = \begin{bmatrix} F_d z \\ 0 \end{bmatrix} \qquad \forall z \in H. \tag{5.93}$$

Next, we show that operator A_c is an infinitesimal generator of a strongly continuous semigroup on Hilbert space H_c. To this end, let $T_c(t)$ denote the following family of linear operators mapping H_c into H_c:

$$T_c(t) \begin{bmatrix} x_d \\ x_{pl} \end{bmatrix} = \begin{bmatrix} T_d(t) x_d \\ T_{pl}(t) x_{pl} \end{bmatrix} \qquad \forall \begin{bmatrix} x_d \\ x_{pl} \end{bmatrix} \in H_c. \tag{5.94}$$

We claim that $T_c(t)$ is a strongly continuous semigroup on H_c, and that operator A_c is the infinitesimal generator of this semigroup.

Using the fact that $T_d(t)$ and $T_{pl}(t)$ are both strongly continuous semigroups, it is easy to show that that $T_c(t)$ is a strongly continuous semigroup on H_c. To show that A_c is the

140

infinitesimal generator of T_c, we take any $\begin{bmatrix} x_d \\ x_{pl} \end{bmatrix}$ in H_c and observe that

$$\lim_{\Delta \to 0^+} \frac{T_c(\Delta)\begin{bmatrix} x_d \\ x_{pl} \end{bmatrix} - \begin{bmatrix} x_d \\ x_{pl} \end{bmatrix}}{\Delta} = \lim_{\Delta \to 0^+} \begin{bmatrix} \frac{T_d(\Delta)x_d - x_d}{\Delta} \\ \frac{T_{pl}(\Delta)x_{pl} - x_{pl}}{\Delta} \end{bmatrix} \tag{5.95}$$

$$= \begin{bmatrix} A_d x_d \\ A_{pl} x_{pl} \end{bmatrix} \tag{5.96}$$

$$= A_c \begin{bmatrix} x_d \\ x_{pl} \end{bmatrix} \qquad \forall \begin{bmatrix} x_d \\ x_{pl} \end{bmatrix} \in H'. \tag{5.97}$$

Observation Process

In a practical active noise cancellation system, the observed signal $y(t)$ is most likely the output of a microphone. A simple model relating the output of a pressure microphone to the acoustic field surrounding it is

$$m(t) = \int_\Omega p(t,\eta)g(\eta)d\eta + v(t), \tag{5.98}$$

where $m(t)$ is the output of the microphone at time t; $p(t,\eta)$ is the pressure field in Ω at time t; $g(\eta)$ is a weighting function in $H^1(\Omega)$; and $v(t)$ is white measurement noise, uncorrelated with all other processes involved. The region of support of $g(\eta)$ is the surface of the diaphragm of the pressure microphone. In this model, the output of the microphone without the measurement noise) is proportional to the weighted integral of the pressure exerted at the diaphragm of the microphone. This model implicitly assumes that the presence of the microphone does not alter the acoustic field in Ω, and that the frequency response of the microphone is constant over those frequencies for which $\int_\Omega p(t,\eta)g(\eta)d\eta$ has significant energy.

141

Recalling that the first component of $x_d(t)$ is the unwanted pressure field at time t, and that the first component of the $x_{pl}(t)$ is the canceling pressure field, we see that $y(t)$ can be expressed in terms of the state-vector $x_c(t)$ as

$$y(t) \;=\; C_c x_c(t) + G_c w(t) \tag{5.99}$$

$$= \; C_c \begin{bmatrix} x_d(t) \\ x_{pl}(t) \end{bmatrix} + G_c w(t), \tag{5.100}$$

where C_c is a linear bounded operator form H' into R^1, and G_c is a bounded linear operator form H into R^1, with

$$G_c G_c^* = Identity, \tag{5.101}$$

and

$$F_c G_c^* = 0. \tag{5.102}$$

Note that eq. (5.101) ensures that the measurement noise $G_c w(t)$ is white, and eq. (5.102) ensures that the measurement noise $G_c w(t)$ is uncorrelated with the process noise $F_c w(t)$.

The explicit operator C_c that results from the microphone model in eq. (5.98) is

$$C_c \begin{bmatrix} x_d \\ x_{pl} \end{bmatrix} = [(x_d^1 + x_p^1), g(\eta)]_{L_2} \qquad \forall \begin{bmatrix} x_d \\ x_{pl} \end{bmatrix} \in H_c, \tag{5.103}$$

where $x_d \in H$ and $x_{pl} \in H'$; and x_d^1 is the first component of x_d and x_{pl}^1 is the first component of x_{pl}. So defined, operator C_c is clearly linear and bounded, since the inner product operation is linear and bounded.

Optimal Feedback Controller

The governing equations of the acoustic system in Fig. 5.2 can be expressed in state-space form as

$$\frac{d}{dt} x_c(t) \;=\; A_c x_c(t) + B_c u(t) + F_c w(t) \tag{5.104}$$

$$y(t) \;=\; C_c x_c(t) + G_c w(t). \tag{5.105}$$

142

The noise cancellation problem is to generate the control signal $u(t)$, based on the observations of $y(t)$, so that the sum of the acoustic energy in Ω and the power expended by the control signal is minimized. Specifically, the cost functional to be minimized is

$$J = \int_0^T E\{\|Q(x_d(t) + x_p(t))\|_H^2\}dt + \int_0^T E\{\|u(t)\|^2\}dt, \qquad (5.106)$$

where Q is bounded linear operator mapping H into H. According to eq. (5.41), $\left\|\frac{\rho_0}{2c^2}(x_d(t) + x_p(t))\right\|_H^2$ is the total acoustic energy in the enclosure at time t. Hence, $\|Q(x_d(t) + x_p(t))\|_H^2$ is the weighted acoustic energy in the enclosure at time t.

The cost function J in eq. (5.106) can be expressed in terms of the state vector $x_c(t)$ as

$$J = \int_0^T E\{\|Q_c x_c(t)\|_{H_c}^2\}dt + \int_0^T E\{\|u(t)\|^2\}dt, \qquad (5.107)$$

where Q_c is a linear bounded operator from H_c into H_c defined as

$$Q_c \begin{bmatrix} x_d \\ x_p \\ x_l \end{bmatrix} = Q(x_d + x_p) \qquad \forall \begin{bmatrix} x_d \\ x_p \\ x_l \end{bmatrix} \in H_c, \qquad (5.108)$$

with $x_d \in H$, $x_p \in H$, and $x_l \in R^n$.

From eq. (5.54) we see that for this problem, a unique optimal control $u_o(t)$ exists and can be expressed in feedback form as

$$u_o(t) = -B_c^* K(t) \hat{x}_c(t),$$

where the linear operator $K(t)$ satisfies

$$\frac{d}{dt}[K(t)x, y] = -[Q_c^* Q_c x, y] - [K(t)x, A_c y] - [K(t)A_c x, y] \qquad (5.109)$$
$$+ [K(t)B_c B_c^* K(t)x, y] \qquad K(T) = 0, \quad \forall x, y \in \mathcal{D}(A_c),$$

and

$$\hat{x}_c(t, w) = E(x_c(t, w)|y(s, w) \quad 0 \le s \le t).$$

Furthermore, $\hat{x}_c(t, w)$ is calculated using a Kalman filter, i.e.

$$\frac{d}{dt}\hat{x}_c(t, w) = A_c\hat{x}_c(t, w) + P_f(t)C_c^*(y(t, w) - C_c\hat{x}_c(t, w)) + B_c u(t, w) \qquad \hat{x}_c(0, w) = 0,$$

143

where

$$P_f(t) = E((x_c(t, w) - \hat{x}_c(t, w))(x_c(t, w) - \hat{x}_c(t, w))^*),$$

and $P_f(t)$ satisfies

$$
\begin{aligned}
\frac{d}{dt}[P_f(t)x, y] &= [P_f(t)A_c^*x, y] + [P_f(t)x, A_c^*y] + [F_cF_c^*x, y] \\
&\quad -[P_f(t)C_c^*C_cP_f(t)x, y] \qquad P_f(0) = \Lambda \quad x, y \in \mathcal{D}(A_c^*).
\end{aligned}
\tag{5.110}
$$

An identical derivation can be carried out for the case in which the boundary Γ of the enclosure is purely absorbing, i.e. eq. (5.16) is satisfied at the boundary in Fig. 5.2.

5.7 Summary

The distributed active noise cancellation problem was studied in this chapter. We focused on minimizing the total acoustic energy in an enclosure. To minimize this acoustic energy, the interaction between the primary field and the canceling field at every point in the enclosure needed to be considered. The acoustic wave equation was therefore utilized to model this interaction. The resulting governing equations for the acoustic system were put in the frame work of infinite-dimensional stochastic linear systems. Using this frame work, we then derived an optimal feedback controller for attenuating the total acoustic energy in the enclosure.

The main contribution of this chapter is showing concretely that under certain assumptions the distributed ANC problem can be put in the frame work of distributed linear control. Based on physical models for the canceling loudspeaker and the acoustic field in the enclosure, a set of coupled ordinary and partial differential equations relating the control signal to the canceling field were derived. We then showed that these equations can be represented as abstract differential equations in an appropriate Hilbert space. This representation was derived by proving that certain operators were infinitesimal generators of strongly continuous semigroups.

The primary field was modeled as the state of an infinite-dimensional linear system driven by a white process. The residual field in the enclosure was modeled as the sum of this primary field and the canceling field. Based on a physical model for the measurement

144

microphone, we showed that the microphone output could be expressed as a linear function of the residual field.

We showed that the distributed active noise cancellation problem in an enclosure with perfectly reflecting or perfectly absorbing boundaries can be put in the above frame work, provided that the canceling loudspeaker is modeled as a collection of point sources of pressure, and the microphones are modeled by eq. (5.98). In our formulation, the control signal is pointwise and is used to drive an ordinary loudspeaker, and the observed signal is pointwise and is obtained as the output of an ordinary pressure microphone. Specifically, our formulation was used to find the feedback controller that minimizes the distributed performance criterion J that was defined in eq. (5.106). More precisely, this performance criterion involves the sum of the acoustic energy in the enclosure and the power expended by the control signal.

The formulation in this chapter is the first mathematically rigorous formulation of the distributed active noise cancellation problem known to us. However, this formulation needs to be extended to the case in which the boundary conditions are more general.

Chapter 6

Conclusions and Future Directions

This thesis dealt with the design of optimal feedback controllers for active noise cancellation. We applied optimal feedback control theory and considered pointwise noise cancellation as well as distributed noise cancellation. By formulating the ANC problem as an optimal feedback control problem, we developed a single approach for designing controllers for pointwise and distributed active noise cancellation.

The key strategy is to model the residual signal/field as the sum of the outputs of two linear systems. The unwanted noise signal/field is modeled as the output of a linear system driven by a white process. Similarly, the canceling signal is modeled as the output of another linear system driven by the control signal. Finally, the residual signal/field is modeled as the sum of the outputs of these two linear systems. These two models are combined to represent the overall acoustic system as a single linear system driven by a control process and a white noise process. We showed that the optimal control signal is a linear feedback of the estimated states of this overall system. These state estimates are computed using a Kalman filter, and the feedback gain is obtained from a Riccati equation. Note that in the pointwise case and the distributed case, the control signal is *not* distributed and is used to drive an ordinary loudspeaker. Moreover, in both these cases, the control signal is generated based on the outputs of ordinary microphones.

While we focused specifically on the acoustic noise cancellation problem, the results developed in this thesis can be applied to other active cancellation problems. Vibration

146

control is an example of a non-acoustic problem to which our results can be applied.

6.1   Summary and Contributions

We started by developing an optimal feedback controller for attenuating the acoustic pressure at the locations of a finite number of microphones. The error microphones are placed at those locations where noise attenuation is desired, and the canceling loudspeaker is placed in the vicinity of these microphones. We model the residual signals at the microphones as the sum of the outputs of two finite-dimensional linear systems. The unwanted noise signals at the microphones are modeled as the outputs of a multiple-input/multiple-output finite-dimensional linear system driven by a white process. This finite-dimensional system provides a compact representation for the second-order statistics of the unwanted signals. Similarly, the canceling signals at the microphones are modeled as the outputs of a single-input/multiple-output finite-dimensional linear system driven by the control signal. This linear system corresponds to the matrix of transfer functions from the input to the loudspeaker to the canceling signals at microphones. Finally, the residual signals at the microphones are modeled as the sum of the outputs of these two finite-dimensional linear systems. These two linear system models are then combined to represent the overall acoustic system as a single finite-dimensional linear system driven by a control process and a white noise process.

We showed that a certain class of performance criterion involving the mean-square sum of the residual signals and the mean-square sum of the control signal could be expressed as a quadratic form in the state of the overall system and in the control signal. The control signal that minimizes this performance criterion is found to be a linear feedback of the estimated states of the overall system. These state estimates can be computed using a Kalman filter, and the control gain can be computed by iterating a Riccati matrix equation. We also showed that the same formulation can be used to minimize the frequency weighted power of the residual signal. Such a frequency dependent weighting might be important because of the difference in the sensitivity of the ear to sounds at different frequencies.

Our pointwise feedback controller has a few important advantages compared to existing feedback controllers for ANC. Recall that analog feedback controllers (e.g., [51]) do not take advantage of the statistics of the noise. However, our feedback controller takes full advantage of the statistical characteristics of the noise. The discrete-time single-microphone feedback controller proposed by Graupe [26] can not be used in the case that the plant model contains one or more delays. However, the feedback controller developed in this thesis can be used in this case. Furthermore, the feedback controller developed in this thesis can be used in the single-microphone as well as multiple-microphone configuration.

Our pointwise feedback controller achieves the same acoustic objective as the existing multiple-channel feedforward controllers for ANC, without the problems associated with acquiring an appropriate reference signal that is commonly experienced by feedforward controllers.

A few experiments were presented to illustrate the performance of the optimal pointwise feedback controller in the context of aircraft noise. The optimal feedback controller outperformed the feedforward controller proposed by Burgess by about 10dB. Similarly, the hardware implementation of the optimal feedback controller outperformed an analog feedback controller by about 9dB.

In the context of distributed active noise cancellation, we focused on developing an optimal feedback controller for attenuating the total acoustic energy in an enclosure. The distributed ANC system utilizes one microphone to monitor the residual field and one loudspeaker to generate the canceling field. The control signal driving the loudspeaker must be generated based solely on the microphone output. To minimize the total acoustic energy in the enclosure, the interaction between the canceling field and the unwanted noise field at every point in the enclosure needs to be considered.

The residual field in the enclosure is modeled as the sum of the outputs of two infinite-dimensional linear systems. The unwanted noise field is modeled as the output of an infinite-dimensional linear system driven by a white process. This infinite-dimensional system provides a compact representation for the second-order statistics of the unwanted noise field. Similarly, the canceling field in the enclosure is modeled as the output of another infinite-dimensional linear system driven by the input to the loudspeaker. We showed that this model can accommodate the finite-dimensional dynamics of the loudspeaker as well as the infinite-dimensional acoustical dynamics of the enclosure, provided that the loudspeaker is modeled as a collection of point sources of pressure and the boundary surrounding the enclosure is perfectly absorbing or perfectly reflecting. This was shown by proving that a certain operate was an infinitesimal generator of a strongly continuous semigroup.

We then illustrated that the linear system representation of the unwanted noise field and the linear system representation of the canceling field can be combined to represent the overall acoustic system as a single infinite-dimensional linear system driven by a control process and a white noise process. Again, this was illustrated by proving that a certain operator was an infinitesimal generator of a strongly continuous semigroup.

Next, we showed that the microphone output at time "t" can be expressed as a linear functional of the state of the overall system at time "t". This was shown based on a physical model for a pressure microphone suggested in [6].

Finally, we considered a class of performance criterion involving the sum of the total acoustic energy in the enclosure and the power expended by the control signal. We demonstrated that this performance criterion can be expressed as the expected value of a quadratic form in the state of the overall system and in the control signal. The control input that minimizes this performance criterion is a linear feedback of the estimated states of the overall system. These state estimates can be computed using a Kalman filter, and the feedback gain can be computed form a Riccati operator equation.

Our results for the distributed ANC problem are important because they are the first mathematically rigorous formulation of the distributed ANC problem. Furthermore, it has been demonstrated in numerical studies that attenuation of noise at a finite number of spatial points can result in amplification of the noise at other points [15].

Comparing our formulation of the pointwise ANC problem to our formulation of the distributed ANC problem, we see the fundamental importance of being able to represent the acoustic system as a linear system driven by a control process and a white noise process. Similarly, to use this formulation, it is essential that the performance criterion be an expected value of a quadratic form in the state of this linear system and in the control

149

process. In the pointwise case, it was straight forward to represent the acoustic system in this form; however, in the distributed case, rather sophisticated mathematical machinery was needed to represent the overall acoustic system in this form.

Several recursive/adaptive algorithms for modeling the unwanted noise at a single microphone were developed in Chapter 4. We assumed that the unwanted noise at the microphone is the output of an all-pole transfer function driven by a white process and develop a recursive/adaptive algorithm for estimating the parameters of this transfer function based on the measurements made by the microphone. An adaptive feedback ANC system can be obtained by combining the pointwise optimal control law with these recursive/adaptive algorithms for modeling the unwanted noise. Note that the resulting adaptive controller only adapts to changes in the noise statistics and not to changes in the plant. Although we focused specifically on developing estimation algorithms for modeling the unwanted noise in the ANC problem, the algorithms presented in Chapter 4 can be applied to the more general problem of identifying non-stationary autoregressive (AR) processes embedded in white noise.

6.2   Future Directions

We foresee several immediate branches down which the work in this thesis may continue.

Developing adaptive optimal feedback controllers for pointwise ANC is a potentially promising area for future work. Our derivation of the optimal pointwise feedback controller was based on the assumption that the noise model and the plan model were known. However, in some applications both the plant characteristics and the unwanted noise characteristics might be unknown or changing. By properly combining an adaptive parameter estimation algorithm with the control law developed in Chapter 3, an adaptive optimal feedback controller for ANC can be obtained. Adaptive optimal control is a mature field, and many of the results developed in this field should be directly applicable here.

Our results for the distributed ANC problem were derived for enclosures with perfectly reflecting or perfectly absorbing boundaries. Extending these results so that they apply to enclosures with more general boundary conditions would be very useful. In the more general case, the normal component of particle velocity at the boundary and the pressure at the boundary are related through an ordinary differential equation. It needs to be shown that operator

$$\begin{bmatrix} 0 & -c\,\nabla. \\ -c\,\nabla & 0 \end{bmatrix} \tag{6.1}$$

, with domain consistent with the above boundary conditions, is an infinitesimal generator of a strongly continuous semigroup in an appropriate Hilbert space.

A computationally efficient method for implementing the optimal control law in the distributed case is lacking. More generally, we feel that a careful study of the computational aspects of the control law in the distributed case needs to be carried out.

Bibliography

[1] R.A. Adams, *Sobolev Spaces*, N.Y., Academic Press, 1975.

[2] K. Astrom and B. Wittenmark, *Computer Controlled Systems: Theory and Design*, Englewood Cliffs, N.J., Prentice-Hall 1984.

[3] M. Athans and P. Falb, *Optimal Control: an introduction to the theory and applications*, N.Y., McGraw-Hill, 1966.

[4] O.L. Angevine, "Active Acoustic Attenuation of Transformer Noise," *Proc. Internoise81*, 1981.

[5] A.V. Balakrishnan, *Applied Functional Analysis*, New York: Springer-Verlag, 1981.

[6] Leo. L. Beranek, *Acoustics*, McGraw-Hill, 1954.

[7] M. Berengier and A. Roure, "Broad-Band Active Sound Absorption in a Duct Carrying Uniformly Flowing Fluid," *Journal of Sound and Vibration* **68**, 1980.

[8] G. Box and G. Jenkins, *Time Series Analysis, Forecasting and Control*, San Francisco, CA: Holden Day, 1970.

[9] A.J. Bullmore, P.A. Nelson, A.R.D. Curtis, and S.J. Elliot, "The Active Minimization of Harmonic Enclosed Sound Fields, Part II," *Journal of Sound and Vibration* **117**, 1987.

[10] A. Burgess, "Active adaptive sound control in a duct: A computer simulation" *Journal of the Acoustical Society of America*, Vol 70, No. 3, pp. 715–726, September, 1981.

[11] G. Canevet, "Active Sound Absorption in an Air Conditioning Duct," *Journal of Sound and Vibration* **58**, 1987.

[12] C. Carme, *Absorption acoustique active dans les cavites*, Ph.D. Thesis, Universite D'Aix-Marseille II, 1987.

[13] G.B. Chaplin, "Method and Apparatus for canceling Vibration," *U.S. Patent Number* 4,489,441 December 18, 1984

[14] A.R.D. Curtis, P.A. Nelson, S.J. Elliot, and A.J. Bulmore, "Active Suppression of Acoustic Resonance," *Journal of Acoustical Society of America,* 1987.

[15] A. David and S.J. Elliot, "Numerical Studies of Actively Generated Quiet Zones," *Applied Acoustics,* 1993.

[16] A.P. Dowling and J.E. Ffowcs Williams, *Sound and Sources of Sound,* New York: John Wiley and Sons

[17] J.E. Efowcs Williams, I. Roebuk, and C. Ross, "Anti Phase Noise Reduction," *Physical Technology, 16.*

[18] S.J. Elliot, I.M. Stothers, and P.A. Nelson, " A Multiple Error LMS Algorithm and Its Application to the Active Control of Sound and Vibration," *IEEE Trans. ASSP,* VOL. ASSP-35, 1987.

[19] S.J. Elliot, A.R.D. Curtis, A.J. Bullmore, and P.A. Nelson, "The Active Minimization of Harmonic Enclosed Sound Fields, Part III," *Journal of Sound and Vibration* **117**, 1987.

[20] S.J. Elliot, C.C. Boucher, and P.A. Nelson, "The Behavior of a Multiple Channel Active Control System," *IEEE Trans. Signal Processing,* VOL. 40, NO. 5, 1992.

[21] S.J. Elliot and P.A. Nelson, "Algorithm for Multichannel LMS Adaptive Filtering," *Electronics Letters,* VOL. 21, NO. 21, 1985.

[22] S.J. Elliot, P. Joseph, A.J. Bullmore, and P.A. Nelson, "Active Cancellation of a Point Pure Tone Diffuse Sound Field," *Journal of Sound and Vibration* **120**, 1988.

[23] L.J. Eriksson, M.C. Allie, and C.D. Bremigan, "Active Noise Control Using Adaptive Digital Signal Processing," *Proc. ICASSP*, New York, 1988, pp. 2594-2597.

[24] Franklin, Powell, and Workman, *Digital Control of Dynamic Systems*, Addison-Welsely, 1990.

[25] D.C. Gilliland, "Sequential Compound Estimation," *Annual Mathematical Statistics*, Vol. 39, No. 6, pp 1890-1904, 1968.

[26] D. Graupe and A. Efron, "An Output-Whitening Approach to Adaptive Noise Cancellation," *IEEE Trans. Circuits and Systems*, Vol. 38, No. 11, November 1991.

[27] D. Guicking, *Active Noise and Vibration Control: Reference Bibliography*, 1991, University of Gottingen, West Germany.

[28] M.J.M. Jessel, "La question des absorbeurs activs," *Revue d'acoust* 37, 1972.

[29] M.J.M. Jessel and O.L. Angevine, "Active Acoustic Attenuation of a Complex Noise Source," *Inter Noise80*, December 1980.

[30] M.J.M Jessel and G.A. Mangiante, "Active Sound Absorbers in an Air Duct," *Journal of Sound and Vibration* **23**, 1972.

[31] A.J. Kempton, "The Ambiguity of Acoustic Sources-a Possibility for Active Control," *Journal of Sound and Vibration* **48**, 1976.

[32] K. Kido, "Reduction of Noise by Use of Additional Sources," *Proc. Internoise75*, 1975.

[33] E. Kreyszig, *Introductory Functional Analysis with Applications*, New York: Wiley and Sons, 1978.

[34] R.C. Lee, *Optimal Estimation, Identification, and Control*, MIT Press, Cambridge, 1964.

[35] J.S. Meditch, *Stochastic Optimal Linear Estimation and Control*, McGraw-Hill, N.Y. 1969.

[36] P.A. Nelson, A.R.D. Curtis, S.J. Elliot, and A.J. Bullmore, "The Active Minimization of Harmonic Enclosed Sound Fields, Part I," *Journal of Sound and Vibration* **117**, 1987.

[37] Nelson, Curtis, Elliot, and Bullmore, "The Minimum Power Output of Free Field Point Sources and the Active Control of Sound," *Journal of Sound and Vibration* **117**, 1987.

[38] Y. Nogami, "The k-extended set-compound estimation problem in a nonregular family of distribution over $[\theta, \theta + 1]$," *Annal Institute of Statistical Mathematics*, Vol. 31A, pp. 169-176, 1979.

[39] J.T. Oden, and J.N. Reddy, *An Introduction to the Mathematical Theory of Finite Elements*, Wiley and Sons, 1976.

[40] A. Oppenheim, E. Weinstein, K. Zangi, M. Feder, and D. Gauger, "Single Sensor Active Noise Cancellation Based on the EM Algorithm," *Proc. ICASSP*, San Francisco, 1992, Vol. I, pp. 277-280.

[41] J.H.B. Poole and H.G. Leventhall, "An Experimental Study of Swinbanks' Method of Active Noise Attenuation of Sound in Ducts," *Journal of Sound and Vibration* **49**, 1976.

[42] H. Robbins, "Asymptotically subminimax solutions of compound statistical decision problems", *Proceedings of the 2nd Berkeley Symposium on Mathematical Statistical Problems*, pp. 131–148, 1951.

[43] H. Robbins and S. Monro, "A stochastic approximation method", *Annals of Mathematical Statistics*, 22:400–407, 1951.

[44] C. F. Ross, "A Demonstration of Active Control of Broadband Sound," *Journal of Sound and Vibration* **74**, 1981.

[45] J. Van Ryzin, "The sequential compound decision problem with $m \times n$ finite loss matrix," *Annual Mathematical Statistics*, Vol. 37, pp. 954–975, 1966.

[46] S.D. Snyder and C.H. Hansen, "Active Noise Control in Ducts: Some Physical Insights," *Journal of Acoustic Society of America*, 1989.

[47] S. D. Snyder and C. H. Hansen, "Design Considerations for Active Noise Control Systems Implementing the Multiple Input, Multiple Output LMS Algorithm," *Journal of Sound and Vibration* **159**, 1992.

[48] S. D. Snyder and C. H. Hansen, "The Influence of Transducer Transfer Function and Acoustic Delays on the Implementation of the LMS Algorithm in Active Noise Control Systems," *Journal of Sound and Vibration* **141**, 1990.

[49] M.A. Swinbanks, "Active Control of Sound Propagation in Long Ducts," *Journal of Sound and Vibration* **27**, 1973.

[50] G.E. Warnaka, L. Poole, and J. Tichy, "Active Attenuator," *US Patent Number* 4,473,906 September 25, 1984.

[51] B. Wheeler, *Voice Communication in the cockpit noise environment- the role of active noise reduction*, Ph.D. Thesis, University of Southampton, England, 1986.

[52] B. Widrow *et al.*, "Adaptive noise canceling: principles and applications," *Proc. IEEE*, Vol. 63, No. 12, December 1975.

[53] B. Widrow and S.D. Stearns, *Adaptive Signal Processing*, Prentice Hall, Englewood Cliffs, NJ 1985.

[54] S.B. Vardeman, "Admissible solutions of k-Extended finite state set and sequence compound decision problems," *Journal of Multivariate Analysis*, Vol. 10, pp 426-441, 1980.

[55] E. Weinstein, A. Oppenheim, K. Zangi, M. Feder, and D. Gauger, "Single-Sensor Active Noise Cancellation," *IEEE Trans. Audio Processing* April 1994.

[56] E. Weinstein, M. Feder, and A. Oppenheim, "Signal Enhancement Using Single and Multiple-Sensor Measurements," MIT-RLE Technical Report No. 560, December 1990.

[57] N. Wiener, *Extrapolation, Interpolation, and Smoothing of Stationary Time Series, with Engineering Applications*, New York: Wiley, 1949.

[58] P.E. Wellstead and M.B. Zarrop, *Self-tuning Systems: Control and Signal Processing*, Chichester, N.Y, Wiley, 1991.

[59] K.C. Zangi, "A New Two Microphone Active Noise Cancellation Algorithm," *Proc. ICASSP 1993*, Minneapolis, Vol. II, pp. 351-354, April 1993.

www.ingramcontent.com/pod-product-compliance
Lightning Source LLC
Chambersburg PA
CBHW082008190326
41458CB00010B/3113